Thomas Buzzard

On Some Forms of Paralysis From Peripheral Neuritis of Gouty, Alcoholic, Diphtheritic, and Other Origin

Thomas Buzzard

On Some Forms of Paralysis From Peripheral Neuritis of Gouty, Alcoholic, Diphtheritic, and Other Origin

ISBN/EAN: 9783744670326

Printed in Europe, USA, Canada, Australia, Japan

Cover: Foto ©berggeist007 / pixelio.de

More available books at **www.hansebooks.com**

ON SOME FORMS

OF

PARALYSIS

FROM

PERIPHERAL NEURITIS;

OF GOUTY, ALCOHOLIC, DIPHTHERITIC, AND OTHER ORIGIN.

THE HARVEIAN LECTURES FOR 1885.

BY

THOMAS BUZZARD, M.D.Lond.

FELLOW OF THE ROYAL COLLEGE OF PHYSICIANS IN LONDON; FELLOW OF KING'S COLLEGE
LONDON; PHYSICIAN TO THE NATIONAL HOSPITAL FOR THE
PARALYSED AND EPILEPTIC.

LONDON
J. & A. CHURCHILL
11, NEW BURLINGTON STREET
1886

PREFACE.

A CONSIDERABLE amount of material which was unavoidably omitted (from considerations of time and space) in the Lectures as they were delivered at the Harveian Society of London, and afterwards published in the 'Lancet,' has been added to the text of the work as it now appears.

June, 1886.

CONTENTS.

LECTURE III.

ON SOME FORMS OF PARALYSIS

FROM

PERIPHERAL NEURITIS:

OF GOUTY, ALCOHOLIC, DIPHTHERITIC, AND OTHER ORIGIN.

LECTURE I.

Introduction—Liability of symptoms due to lesion of peripheral nerves to be referred to central disease— Localised forms of neuritis — Multiple neuritis — Causes of paralysis — Anatomical considerations — Waller's degeneration—Electrical reactions in lesions of nerves — Variation of symptoms in neuritis— Favorable prognosis—The term "neuritis" used provisionally in certain cases—Case of typical neuritis— Cases suggestive of neuritis from gout—Multiple neuritis described by Graves—Multiple neuritis from various causes: alcoholism, diphtheria, syphilis, kakké or beriberi.

MR. PRESIDENT AND GENTLEMEN,—When your Council honoured me with a request to deliver the Harveian Lectures this year, my first care was to see whether there was anything in the Bye-laws of the Society to aid me in the difficult task of selecting a subject upon which to address

1

you.   The only direction that I found was that the lectures should be upon some subject of practical interest in medicine, surgery, or midwifery.   In the choice which I have made, with whatever amount of success it may prove to be carried out, at least I can lay claim to the selection of a subject the practical interest of which it would be difficult to surpass.   The present generation has seen remarkable advances in our knowledge of diseases of the nervous system.   A flood of light has been thrown upon the physiological anatomy and the pathology of the brain and spinal cord, and it is perhaps to some extent in consequence of this comparative concentration upon the central nervous system that the part played by the peripheral nerves in the production of symptoms of disease has not until recently received the amount of attention which it undoubtedly deserves.

The improvement in our knowledge is largely due to the development of microscopical examination.   Science is indebted to the late Lockhart Clarke for the origination of processes of hardening and staining nerve tissue which have now acquired great perfection.   How much of our acquaintance with diseases of the brain and spinal cord we owe to the microscope aided by methods of this kind is well known to all. Within the last few years these methods, improved by experience, have been applied to the examination of peripheral nerve-fibres with the result of disclosing a large amount of previously unsuspected change in the essential elements of the nerve-fibre itself.

It will be my aim in these lectures to show that many forms of paralysis which would at first sight point to organic disease of the central nervous system are in all probability dependent essentially upon changes in the periphery of the cerebro-spinal nerves. The subject is a long one, and the time at my disposal is short. It will be necessary, therefore, to deal with the topic in a somewhat general manner, avoiding, where possible, any wearisome minuteness of detail, and passing over such branches of it as may fairly be considered common knowledge at the present day. Even with these limitations I shall have to ask your forbearance for many a cautious expression and not a few acknowledgments of ignorance. There are indeed necessarily many great gaps in our knowledge of the subject, and numerous points on which it seems difficult to reconcile apparently conflicting circumstances, for it is comparatively young, though growing vigorously from day to day.

The plan which I propose to adopt is first to narrate a case which may be taken as a typical example of a localised peripheral neuritis; then to show, by reference to clinical cases, the probability that neuritis may occur in forms not easily recognisable, owing to the absence of one or more of the characteristic symptoms. The possibility of its frequent occurrence in the gouty diathesis will be considered, and examples furnished. In these circumstances it is in one limb or portion of the body that the manifestation usually takes place, and the idea of a stroke of hemiplegia is readily suggested. Passing from the localised form of paralysis, we shall next consider cases of multiple neuritis—cases in which there is a tendency to rapid and almost universal paralysis. It is in these that the imitation of central nervous disease is apt to be the strongest. We cannot in every case of this disease feel at all certain as to what is the cause of the changes in the peripheral nerves upon which the symptoms depend. But there is increasing evidence to show that many are due to some toxic influence introduced from without. Amongst influences of this kind are to be mentioned alcohol, lead, diphtheria, syphilis, and something which gives rise to the endemic disease of Japan and the shores of Eastern Asia, called Beriberi or Kakké. Paralysis, as my master, the late Dr. Todd, used to teach, is not a disease of itself, but a symptom—an effect due to a cause, which cause itself is not always the essential disease. The paralysis in the cases under consideration is immediately due to the change in the nerve-fibre, but this in its turn depends upon a cause which is sometimes, but not always, recognisable. The diagnosis of this class of cases will next claim our attention, and especially the mode by which they may best be differentiated from diseases of the spinal cord. The prognosis and treatment of the disease will suitably conclude our observations.

A few preliminary remarks may not be out of place.

Although it is desirable that we should understand what we mean when speaking of paralysis it is not necessary to attempt a strict definition of the term. We know that in health the muscles contract in obedience to impulses conveyed to them from the cerebro-spinal centre. When we speak of paralysis we mean that this process fails to be efficiently performed. The imperfection may be extremely slight or the function may be completely suspended, but in any case the resulting condition is one of paralysis.

The cause of failure may lie in the centre

itself, or in the peripheral nerves by which the so-called "motor impulses" should be carried to the muscles.

On the sensory side again we see analogous deviations from normal functioning. In a state of health afferent impulses from a sentient surface are propagated along certain fibres in the peripheral nerves up to cells in the central nervous system, giving rise in them to changes which are represented in consciousness as sensations. Lesion of the peripheral nerve-fibres, equally with lesion of the central nerve-cells, will interfere more or less completely with the normal function, and loss or some disorder of sensation will be the result.

On the present occasion it will be understood that I shall confine myself to some of those forms of paralysis which are essentially dependent upon lesion of the peripheral nerve-fibres, and in considering these such as are the result - of traumatic agency will be excluded.

It will be sufficient to remind you, in a very few words, of the anatomical constitution and arrangement of the peripheral nerves. The cerebro-spinal nerves are individually connected by one or more roots with the brain or spinal cord. Each spinal nerve arises by two roots, an anterior, which contains the motor fibres of the nerve, and a posterior, which contains the sensory fibres. On the latter root is a ganglion, and immediately beyond this the two roots unite into a single nerve-trunk, which quits the spinal canal by the intervertebral foramen. The nerve thus formed is spoken of as a " mixed " nerve, *i. e.* it is endowed with both motor and sensory functions. Outside the canal it divides into two branches, both containing motor and sensory fibres, which are distributed to the anterior and posterior parts of the body; the anterior, which is the larger, supplying the limbs and the front of the body generally.

The trunk of a spinal nerve is composed of nerve-fibres, side by side, in form of a bundle, the fibres being individually separated by connective tissue, and the bundle itself provided with a special sheath of the same. This sheath is lamellated, and contains lymph-channels. The bundles of nerve-fibres thus constituted are connected and held together in their turn (so as to form a trunk) by connective tissue, in which are contained lymphatics, blood-vessels, and fat-cells. The individual nerve-fibre (bundles of which go to make up a nerve-trunk) consists of a primitive sheath or membranous tube, containing the white substance of Schwann. This probably serves to insulate the essential constituent of the nerve, the axis-cylinder which occupies, as the term given to it implies, the centre of the tube.

The important observation made by Waller is requisite to be borne in mind in connection with our subject. It will be remembered that this distinguished physiologist found that when the anterior root of a mixed nerve is divided the nerve-fibres thus separated from the spinal cord degenerate to their remotest periphery in the course of a few weeks. If, however, the posterior root be cut through between the cord and the ganglion upon the root it is now not the periphery but the portion of the nerve attached to the cord which degenerates. The ganglionic cells of the anterior horns appear therefore to act as trophic centres for the motor fibres, whilst the spinal ganglia would seem to serve the same office for the sensory fibres. Waller formulated the general law that nerve-fibres degenerate when separated from their trophic centres. The term Wallerian degeneration is commonly applied to the changes taking place in nerve-fibres in these circumstances.

A very brief reference is necessary to the changes of reaction to electrical currents which occur in a muscle when the nerve distributed to it is from some cause more or less completely disabled in its power of conduction.

In mild cases no change is to be noted. The muscle reacts to induced currents of the same strength as those which bring about contraction in a similarly placed muscle supplied by a healthy nerve.

In more severe cases, as shown by Erb, whilst the muscle continues to respond to induced currents its excitability to a slowly interrupted galvanic current may be increased beyond what is natural in health.

Where the condition is one of still greater severity, the muscle, after showing for a few hours or days increased excitability to electric currents, rather rapidly loses its power of response to induced currents, so that in the course of two or three weeks no contraction is elicited by the application of the strongest currents. At the same time slow interruptions of a galvanic current of a strength quite insufficient to visibly affect healthy muscle determine very definite contractions. These contractions differ from those of healthy muscular tissue in being slower and, so to speak, of a more lazy character. It is now found too that the contraction consequent upon closure of the galvanic circuit with the positive pole upon the muscle takes place when a much more feeble current is employed than that which is required to bring this about in health. In a normal state, as is well known, the contraction most easily and earliest brought about is that produced when the galvanic circuit is closed with the negative pole upon the muscle. Now, however, the condition is reversed, and it is when the circuit is closed with the positive pole on the muscle that we get the earliest result, *i.e.* we obtain a contraction with the least number of cells. In the course of several months a gradual recovery of power over the affected muscles may be expected to take place when the lesion does not cause permanent change. It often happens that the voluntary power of contracting the muscles is regained before they will respond to the induced current but after they have ceased to be abnormally excited by galvanism. When this occurs it is probably only a question of time as to when faradism will be again able to exert its ordinary exciting influence.

These are the main features of the "reaction of degeneration," a term introduced by Erb, to which the symbol RD has been conveniently applied in this country. This symbol will sometimes be employed in these lectures to avoid the wearisome repetition either of the long phrase, or of the description of the condition to which it has been applied.

I have purposely referred first of all to the changes of reaction in the muscles, because it is these which immediately strike the eye when we proceed in the ordinary way to examine a patient with electrical currents. With more propriety, perhaps, from a physiological and pathological point of view I should have first referred to the changes to be noted in the nerve-trunks on their way to furnish the intramuscular fibres. These, however, are more difficult to test. As a rule, their reaction is the same as that of the muscles to which they are supplied, except in the partial form of RD, when the response to currents remains normal in the nerve whilst qualitative and quantitative changes are manifest in the muscle. But the electrical reactions of nerves are from various causes subject to apparent anomalies which we cannot as yet account for, and which interfere with the exactness of the information to be derived from their behaviour. We may, however, recognise changes in the irritability of a nerve to electric currents, whether in the direction of plus or minus, and these will at least indicate that the nerve is the seat of *some* lesion or organic change—some alteration probably in its nutrition. But neither the nature nor degree of the lesion is always to be inferred from the changes of excitability to electric currents displayed by the trunk of a nerve.

In general terms it may be said, as it is well put by Ross, " The state of the nutrition of a nerve has great effect upon its irritability. If the nutrition is wholly arrested the irritability disappears. But a nerve whose nutrition is merely defective discharges its energy more readily than one whose nutrition is perfect."

Where the trophic influence of the spinal cord is obstructed by such a lesion of the motor nerve as brings about the reaction of degeneration, it is usual to find that the muscles so affected are apt to become rapidly atrophied. This atrophy, like the reaction of degeneration, is frequently but a temporary condition which is quite capable of complete recovery. Where, however, the nerve has undergone irreparably destructive changes, the atrophy of the muscle to which it is supplied is permanent.

With this brief reference to certain points which may usefully be borne in mind, I will proceed to relate a case which has not long since fallen under my observation, and will serve as a convenient introduction to our subject. It is a typical case of neuritis affecting certain branches of the brachial plexus, and occasioning local paralysis, exquisite pains, hyperalgesia, muscular atrophy, abolished or diminished electrical excitability, and trophic changes in the skin. In this example the collection of symptoms points conclusively to lesion of peripheral nerves, and if neuritis were characterised like this in all instances there would be comparatively little difficulty in its diagnosis. But it is an important fact that we may have paralysis from neuritis—frequently of a progressive and multiple form—in which one or more of the symptoms that I have just enumerated may be entirely wanting. The absence or want of prominence of such symptoms tends very often to obscure the diagnosis so much that the condition stands the chance of being referred to central instead of peripheral nerve changes. Now, this alone is an important point, because without correctness in diagnosis our modes of dealing with disease must necessarily be more or less of a haphazard character. But more than this—the forms of disease to which I am going to draw your attention, whilst often presenting the most alarming features, may show themselves highly responsive to treatment, and issue in recovery

to an extent which could hardly be expected
from the grave aspect often presented at the
outset of the attack.   Unfortunately, a considerable part of the field of disease dependent
upon lesion of the nervous system is occupied
by more or less intractable disorder.   It is
therefore with peculiar interest and satisfaction
that I find myself able to discuss, in these
lectures, forms of disease in which a favorable
prognosis is very often indeed justified by the
result.

A word is necessary as to the employment of
the term "neuritis."   It is common at the
present time to distinguish two forms of
neuritis : interstitial neuritis, in which the connective tissue of the nerve is the primary seat
of inflammatory changes, the essential element
being secondarily affected; and parenchymatous neuritis, in which there is destruction of
the essential element of the nerve-fibres, with
but little, or even, perhaps, no recognisable
alteration in the interstitial tissue.   It is open
to discussion how far we are justified in applying a term which suggests inflammation to the
latter of these two processes.   The change
which takes place in the nerve-fibre is of the
character of degenerative atrophy, and is comparable with that which occurs below the point

of section when a motor nerve is artificially divided. These two forms of lesion, interstitial and parenchymatous neuritis, call to mind the changes which are observed to take place in the nerve-fibres constituting the posterior columns of the spinal cord in cases of so-called sclerosis. In some of these it is manifest that the investing tissue of the fibre has been primarily affected, the essential nerve-elements suffering secondarily; whilst in others, as is commonly though not universally accepted, the nervous elements are the first to undergo change which spreads to the interstitial tissue. Whether in the latter case there is justification for the employment of a term implying inflammatory action is not quite certain. The question, both as regards the changes in the nerve-fibres which run in the spinal cord, and in those which take their course in a nerve-trunk, must be acknowledged to be still in an unsettled state. In applying, therefore, the term "neuritis" to changes in the essential nerve-elements of a fibre when these are to all appearance primarily as well as when they are secondarily produced, it will be understood that I suspend my judgment as to the propriety of the parenchymatous form being considered as certainly of inflammatory character, and make use of the expression as a convenient one which has the advantage of being generally recognised.   The following example of neuritis was in all probability of interstitial character :

A single woman, aged twenty-four, was sent to me from the country on February 20th, 1883, suffering from loss of power in the right hand, with agonising pain. Her arm was in a sling, the hand covered up with cotton wool, and she jealously watched the limb to guard it from the slightest accidental touch, so exquisite was the tenderness. The right hand and forearm had a soddened, puffy, help-less appearance, with swollen fingers and purplish dis-colouration of the skin in patches, which here and there looked glossy. Her immediate illness had commenced in the preceding August (six months previously) with pain and swelling in the middle finger, which gradually extended to the others, and for some months past her hand had been quite useless. The pain was so constant and severe that she could scarcely ever get sleep at night. She looked extremely ill. It seemed that her business was to help in the household of her father. She had lost her mother from cancer of the liver. There was no rheumatism in the family, but she had been considered to be weak in the chest. A brother was supposed to be consumptive, and a sister had cough. She herself had suffered much from so-called "rheumatism" in the knee and left hand. Nothing wrong was to be found on examining her chest, and the ophthalmoscope showed no change in the fundus oculi. There was not the least reason to suspect either alcoholism or syphilis. Her temperature was 100° F. On examination, it was seen that the power of extending the wrist was moderately good, but flexion of it could not be performed. There was slight power of flexing the last joint of each finger, and an equally slight power of extending it, and this applied also to the last joint of the thumb. There appeared to be no power in the intrinsic muscles of the thumb and fingers. Examined electrically, the thenar muscles did not respond to either form of electric excitation; but the muscles of the back and front of the forearm were excitable by induced currents, though only when a considerable strength was employed. A detailed examination was not possible, owing to the exquisite sensitiveness of the limb. Warmth was felt as well by the right hand as by the left, but cold was felt best on the left (unaffected) hand. A hair drawn over the skin of the left hand was felt better than on the right. The patient was forced to keep the limb covered up, as the air would start pain, and conveyed a burning, smarting sensation. There was a more or less constant feeling of numbness in the fingers.

It was very difficult to obtain a clear history of the commencement of her illness. Her mother, it seemed, had died in March, and patient had nursed her night and day for a year, never having a really good night's rest. She fell ill herself after her mother's death, but at first there was nothing wrong with the hands. But about three or four weeks afterwards there came a patch of inflammation on the middle finger of the *left* hand which broke after being poulticed, and healed up. It had never kept her awake. Soon after this she had an inflamed throat, with fever and a swelling under the jaw which did not discharge. She was confined to her bed during fourteen days. In August, the middle finger of the *right* hand began to get red, and the inflammation spread in turn to the index, ring, and little finger, thence to the thumb and hand, and then to the wrist of this side. At the same time pain began to be felt on movement, with ten-

2

derness of skin, which had increased and continued ever since. She had been especially getting worse since Christmas.

I will not weary you by any detailed account of her progress.

By way of treatment, the limb was supported by a splint, ice and small flying blisters being applied; nourishing diet was ordered, and opium administered internally. The symptoms, however, continued without any material change, except that in June, after placing her arm in hot water, it became "spotted" all over, as she described it (for I did not see this), and blisters formed over her fingers. The blisters, apparently of the nature of pemphigus, discharged, and became covered with crusts, which remained when I saw her in July. I lost account of her after this, but have since learned the sequel, which is sad. She continued to suffer as described during the winter of 1883-84, and in the early spring of 1884 was attacked with acute melancholia with strong suicidal intent, and was confined for six months in an asylum. As described by the superintendent of the asylum, "her arm at the time of admission was somewhat smaller than the other, with diminished mobility and considerable pain. There was some discolouration of the skin." The arm was supported for a time, and by the end of April it appeared to have recovered its size, and she could use it a little. The pain was not constant, but warmth caused a "burning feeling," and cold "painful rheumatic sensations." By September she could use the right hand as well as the left. Morphia was administered to her in the asylum until her melancholic symptoms were considerably improved. In October she was discharged, recovered both in her mind and also in her arm, which she used freely and without pain. The only complaint when last I heard of her was that in cold weather the arm ached.

This case was one of a rare and important character, for, without any history of violence, the symptoms bore a close resemblance to those so graphically described by Weir Mitchell as the result of gunshot injury to nerves. It was marked by paralysis, loss of or diminished electrical reaction of muscles, agonising pains of lancinating character, constant burning sensation, exquisite hyperæsthesia of the skin and probably also of the muscles, but it was not possible to separate absolutely the two conditions. Besides these symptoms, pointing to lesion of motor and sensory fibres, there were others which appeared to indicate that the vaso-motor fibres were also involved. These were the sodden, œdematous look of the limb, the patches of glossy skin (a symptom first described by Paget), the purple discolouration, and the bullous eruption which left behind it adherent crusts.

It will have been noted that the lesion of the right arm had been preceded by some kind of local inflammatory condition of the middle finger of the left hand, which had been followed by inflamed throat and swelling under the jaw. The tendency to symmetrical affection of the opposite

side in cases of neuritis has been referred to by
Weir Mitchell and others, but whether we are
to see in the symptoms described any connection of this kind, I am quite unprepared to
say.

The case was clearly one of neuritis, the
essential cause of which was, however, not
evident. In process of time all the formidable
symptoms subsided and the patient recovered.
There must therefore have been regeneration
of nerve-fibres. Whilst there was paralysis of
the muscles of the forearm as well as of the
hand, it is to be noted that those of the latter
were much more severely affected than the
former. There was no power of voluntarily
moving the thumb, except at its last joint, and
the thenar and interosseous muscles showed no
response to either form of electric excitation.
The power of moving the long muscles on the
back and front of the forearm was not altogether lost, and they proved to be excitable by
induced currents, though of a much greater
strength than is needed by healthy muscles.
The effect of the lesion then became more and
more severe as the periphery was approached.
I would draw especial attention to this point,
because, as we shall see, it plays an important
part in the diagnosis of other forms of paralysis

which are not so palpably dependent upon lesion of the trunks of nerves.

The trophic disorder of the skin which here occurred in so marked a form is by no means always present in cases of undoubted neuritis. And just as this may be absent, so other of the symptoms may be wanting. We are not yet in a position to explain this important fact, but there would seem to be no doubt that sometimes the motor, at other times the sensory, and perhaps, on the whole, least commonly the vaso-motor fibres, bear the brunt of the attack, with a corresponding contrast in the symptoms. To gain a true idea of the frequency of neuritis it is necessary to remember this fact; we shall otherwise be liable to overlook the true character of affections because they do not display all the characters of a typical neuritis.

Where we are concerned with purely motor nerves we can readily understand that neuritis may occur without giving rise to pain. There can be but little or no doubt that the facial paralysis of peripheral origin which is so common is due to neuritis of some part of the portio dura. The affection is not accompanied by pain, unless, as I have known sometimes happen, the draught of cold air which occasions the neuritis of the seventh nerve sets up at the same time a corresponding change in the fifth. In other purely motor nerves, again, such as the third and sixth, neuritis causes no pain. It may be added that the same immunity is seen when a nerve of special sense—as the optic—is similarly affected. But even in mixed nerves I feel sure that neuritis may occur without causing pain, and I shall have occasion presently to refer to cases which illustrate this point. It has an important bearing upon the question of multiple neuritis, which will be considered later on in these lectures.

It is impossible to affirm positively that we have to do with neuritis, unless, as in the case just read, the aggregation of symptoms leaves us in no doubt. But we are bound to remember that the degree of lesion must be a varying quantity, and that the same symptoms are not to be expected in mild as in severe cases. This may well be illustrated by the example of facial paralysis of peripheral origin, in severe cases of which we find typical reaction of degeneration besides some atrophy of the muscles, whilst in mild instances the electrical reaction is unchanged and the muscles do not waste. There can be no reasonable doubt that we are concerned in each case with a lesion of similar character but of different severity.

The following are illustrations of the resemblance which neuritis may cause to the symptoms of central disease :

A female patient, aged fifty-six, was said to have been suddenly seized one day with loss of power and numbness in the left arm. When she was first seen by her medical attendant these symptoms had passed off to some extent, but she was cold and agitated, with an irregular pulse. There was no paralysis in the face, and the speech was not affected. She walked upstairs with some assistance and remained in bed for about ten days, her symptoms during that time being occasionally severe headache, with loss of appetite and nausea, and on several nights delirious wandering. The urine, which was scanty and of high specific gravity, was loaded with lithates, and contained no albumen. The result of examination of her organs was negative. The left arm was often complained of as weak, heavy, and numb. On inquiry, it appeared that her illness had not really been quite so sudden in its onset as had been thought, but that the loss of power and numbness had been preceded by a great deal of pain in the left arm and shoulder. She now gradually improved, though unequal to the slightest strain or fatigue, and began to get out in the garden. A month after her first attack, whilst out walking, she experienced a sudden loss of power in both legs, which passed off in an hour or so. This did not seem to throw her back very much, but she had several slight threatenings of a return in the following month, now in one limb, now in another. About five weeks after the second attack, whilst a mile from home, she had a severe attack affecting both legs, and had to be brought back in a cab. This gave her a considerable shock, but the paralysis proved to be as transient as before, and a week afterwards, when I saw her for the first time, she was free from any loss of power or abnormal sensation. Examination failed to discover any disease of the nervous centres, or of other viscera. Since that time I understand she has again had a relapse.

This patient had led rather an anxious and chequered life, and the habit of her household was to take stimulants freely. The apparently sudden loss of power in one arm, coupled with numbness of the skin, in a woman fifty-six years of age, is of course highly suggestive of an attack of hemiplegia from some central lesion. But it is to be remarked that the paralysis here had really followed great antecedent pain in the arm and shoulder, an association which is practically conclusive of the peripheral nature of the lesion. The attacks which followed, first in one limb and then in another, clearly point in the same direction. I have no doubt that we have here to deal with neuritis dependent upon either the toxic effect of alcohol or of gouty origin.

Cases of this description are not uncommon, and frequently give rise to a great deal of anxiety. A gouty patient who is past middle age is prone, as we well know, to disease of the blood-vessels and kidneys. An attack of numbness and paralysis in a limb, in these circumstances, is naturally liable to be referred without hesitation to a central lesion : hæmorrhage or thrombosis. That this supposition is far more often than not correct is certainly the case, but every now and then we meet with examples like the one described, which show the necessity of bearing in mind the possibility of a peripheral cause in such an attack. The question of the possibility of gout causing neuritis was referred to by Mr. Hutchinson in the Bowman Lecture at the Ophthalmological Society last year. He adduced some instances which appeared to point to neuritis of the optic nerve originating in gout, as well as others suggesting the occurrence of neuritis in other parts of the nervous system from a similar cause. I have very little doubt that neuritis is not seldom due to the presence of gout; the difficulty of proof is of course extremely great. I cannot lay claim to adduce anything which is absolutely positive upon this point, but some clinical observations, and especially certain electrical examinations which I have made, appear to lend considerable strength to this view.

An old friend of mine, a member of our profession, sent for me a few months since in a state of alarm, having woke up in the morning with numbness in the arms, which was at first slight and affected the side on which he was lying,

the left much more than the right.   In the course of a few
hours it had grown rapidly worse, and was accompanied by
pain in the shoulders.   There was considerable loss of
power in the arms.   I found that he had had a characteristic attack of gout in the ball of one big toe some six years
previously, and more than once had suffered badly from
lumbago.   He had been liable to pains in the shoulders
for many years, and on several occasions had been troubled
with sciatica.   I was able to reassure him unreservedly,
and treatment, directed entirely to his gouty habit, brought
about immediate improvement and complete recovery in no
great time.

It is not certain that we are justified in
applying the term " neuritis " to such cases as
this, with the small amount of evidence of the
pathological condition which is usually to be
obtained.   The contrast as regards the number
and severity of the symptoms with such a case
as that of the young woman which I first
described is very striking ; but it appears to
me that we have only to imagine a like affection
of the nerves, though comparatively of very
slight kind, to explain the symptoms of nerve
lesion observed in gouty patients.   Not only
are both the motor and sensory fibres often
involved in these circumstances, but it frequently happens that there are signs of the
vaso-motor fibres being likewise affected, causing coldness of the extremities and discolouration of the skin.   I lately saw a case in

which, along with pains which were almost universal and there was strong evidence to show were dependent upon a gouty habit, there was an extraordinary amount of œdema of the lower extremities. This extended half-way up each leg. As there was nothing in the state of the heart, kidneys, or liver to explain the dropsical condition, it seems quite possible that the cause of it as well as of the pains may lie in neuritis, which involves both sensory and vaso-motor fibres.

Electrical examination will often give a good deal of support to the view that these are cases of slight neuritis :

A lady, aged fifty-two, complained that her left hand would close during the night, and that she could not get it open again without dreadful pain at the wrist and up the fingers. It would be found icy cold. Some time previously, her left arm, and to a less extent the right arm and the toes of either foot, would "go to sleep." On examination she complained of pain, pricking, and tingling in the thumb and first three fingers, as well as slight numbness and coldness in the toes. Occasionally there would be a dart of pain down the arm and finger. She was a healthy-looking woman, who presented no signs of degenerative changes. Her tongue was clean. She had usually enjoyed good health, except from what she called rheumatism and occasional attacks of indigestion. In these there would be acid risings with bilious vomiting and palpitation of the heart. She was troubled also sometimes with flushes of heat, and according to her account her stomach was easily put out. Her urine was described as being thick. Examination with induced electrical currents showed that the intrinsic muscles of the left thumb were less excitable than those of the right, and, to a marked degree, less excitable than the corresponding muscles of a healthy subject. Inquiry into the history elicited that the patient's father had suffered badly from gout, and that she herself drank a great deal of sherry and occasionally also whisky.

A tradesman, aged forty-seven, complained that the thumb and first two fingers of the left hand had lost grasping power. The skin covering them was more sensitive than on the right side. He could not flex the phalanges of the thumb, but could adduct the member; nor could he flex any of the phalanges of the index and middle fingers except the first. His forearm had become thinner; at its upper part it measured $8\frac{7}{8}$ in. as against $9\frac{3}{8}$ in., the measurement of the right forearm at the same point. The patient had suffered for upwards of three weeks. There had been no pain in the arm, but a numbness and dead feeling over the inside half of it. The fingers, however, were always in a state of painful "pins and needles," a feeling, as the patient himself described it, "exactly like that which occurs after pulling ice about for some time, and then putting the hands near the fire." At the parts of the hand and arm affected by this active numbness there was exaggeration of all forms of sensibility—that of touch, pressure, heat and cold, and pain. I found the faradic reaction of the intrinsic muscles of the thumb defective. With the galvanic current a rheophore over the musculo-spiral nerve a little above the elbow showed that the closure contraction with the anode was equal to that with the kathode, and that there was an opening contraction with the kathode equal to that with the anode. These serial changes apparently pointed to lesion of the nerve. No note was made of the reaction of the median nerve. The nerves implicated in this case were the branches of the median supplied to the palm and first three fingers, as well as those to the opponens, abductor, and flexor brevis pollicis : also the cutaneous branches of the radial distributed to the dorsal surface of the thumb and two outer fingers ; and, in addition, the internal cutaneous from the trunk of the musculo-spiral. Passive extension of the forearm (the hand being in a position of semi-pronation) was strongly resisted by the supinator longus, which was brought out in bold relief by the procedure, and was evidently uninjured.

My notes are, unfortunately, deficient in details regarding the habits of this patient, but I think I am right in saying that they were conducive to gout. There was no history of exposure to cold or pressure. The symptoms are clearly referable to neuritis, which had gone so far as to produce some muscular atrophy, as well as changes of electrical reaction. The case is especially interesting from the absence of pain. In this respect it is comparable with cases of paralysis of the musculo-spiral nerve from cold, first prominently described by Duchenne (de Boulogne), who refers the cause to congestive hyperæmia of irritative character.

A big, healthy-looking man, forty years of age, suffered from numbness in the index finger of the left hand, followed by a pricking or " pins and needles " sensation, and aching pain in the arm, so that he was unable to lie on the left shoulder at night.   There were also, besides this, occasional bursts of darting pain in the arm.   His grasp had lost power.   Examination with the voltaic current showed increase of excitability in the musculo-spiral nerve of the left side; the interruption of a current from two milli-ampères (rheophore on the nerve in the arm above elbow) caused a very painful feeling of an electrical shock to be experienced down to the forefinger on the left side.   A current of five milliampères was required to produce the same effect on the right side.   There were no serial changes. (K S Z > A S Z.)

In cases of this kind you will often find a point on the shoulder where pressure gives exquisite pain.   It lies just inside the inner and upper angle of the scapula, and the pain caused by the pressure there seems to travel down to the hand.   Apparently there is neuritis of the posterior branch of a spinal nerve, the anterior branch of which enters into the formation of the brachial plexus.

A female patient, aged fifty-eight, for two years had suffered from pain in the right thumb of a violent throbbing and darting character.   It would recur daily and last for a longer or shorter time.   After a little while it shot up the radial border of the forearm into the shoulder-joint, collar-bone, and sternum, and later still she got pain in the right eye.   There was a certain periodicity in the pain; she would wake with it; after breakfast there would be some relief, and in the forenoon, although not free from suffering, it would be tolerable.   But about 10.30 every night the pain would strike the thumb and forearm, and give her insupportable suffering till midnight—the attack being accompanied by great flatulence. The arm was said to have lost power.

On electrical examination it was found that at the right median nerve, above the wrist, the excitability to the induced current was increased, whilst with the galvanic current the contraction on opening circuit with the positive pole occurred sooner than the contraction on closure with the negative pole. (A O Z > K S Z.)

This patient drank large quantities of beer besides sherry in the morning and passed much uric acid.

Under treatment there was great reduction of the undue excitability of the nerve, but no marked diminution of the pain. How far the restricted dietary ordered was followed I have no means of knowing.

A man, aged forty-seven, on going to bed one night, felt his right arm become quite numb. He had not been asleep. Next morning he found that he could not throw his coat on as usual. There was no pain. Next day, when I first saw him, there was a feeling of "pins and needles" in the arm, and this, with the powerlessness which still continued, alarmed him very much. On examination it was found that the deltoid muscle was paralysed; he was quite unable to raise the elbow outwards and upwards. When the forearm was extended it could not be passively flexed, showing integrity of the triceps. But when the forearm was flexed on the arm I could forcibly extend it, whether the forearm were in a supine position or semi-pronated, showing weakness of biceps, brachialis anticus and supinator longus. In the right deltoid the reaction to the induced current was slightly reduced, whilst that to slow interruptions of the galvanic current was greatly increased,

contractions occurring to a strength of 1·7 m.a. by the galvanometer (RD).   Five days later it needed a strength of
2·7 m.a. to bring about contractions in the deltoid, and by
this time the other muscles appeared to act normally.   In
another fortnight he used the right deltoid freely, and the
muscle acted to a normal strength of induced current.
This patient had suffered from sciatica, lumbago, and
neuralgic symptoms in various parts at different times, and
had, I think, lived freely.   But I have not details sufficient
to say with any degree of certainty that his affection was
of gouty origin.

In testing electrically for variations of excitability in the nerve-trunks and muscles, it is
important to remember that the resistance
offered by the skin and subjacent tissues may
sometimes vary in the two sides of the body,
although in my experience great care in thor-
oughly wetting both the skin and the rheophores
in hot water, as well as attention to the position
of the latter, minimises this objection.   Dr. de
Watteville has insisted strongly on the necessity of employing a galvanometer in order to
check the results obtained by the application
of currents, and his recommendation should
certainly be followed.   It is necessary to bear
in mind that, although in cases of local paralysis with more or less sensory disturbance,
striking alterations in the electrical excitability
of nerves and muscles may be met with, these

are by no means always present. Nor, as I have said, are the changes when present so constant in character that we can draw any exact inference from them in the present state of our knowledge, except that there is probably *some* tissue change in the substance of the nerve-trunk.

A working man, sixty-nine years of age, applied at the hospital on account of shooting pain, aching, and numbness in the left forearm and hand. He described a kind of tingling from the shoulder to the fingers. After a time, under treatment, he lost the painful sensations, but the hand felt as if asleep, heavy, and big, and the grasp was much weaker than that of the right. Common sensibility was not much affected, but he could not feel a pin to pick it up so well as with the other hand. Cold felt colder, and heat hotter, to the left hand than the right. Although he had never suffered from gout and had not noticed gravel in the urine, the treatment was directed towards the possibility of this disorder. In the course of his attendance he had an unmistakeable attack of acute gout in one of his big toes. About a month afterwards, in July last, he reported his hand and arm much improved, though still to a certain extent numb. In the ulnar and musculospiral nerves electrical examination gave K S Z > A S Z. Towards the end of October, having meanwhile been almost free from discomfort, he again attended with a return of the old symptoms. The thumb and first three fingers were described as feeling quite dead, and the grasp of the hand was very weak. Examination of the median nerve at the bend of the elbow with a galvanic current was now made. The resistance of the tissues over the nerve in

3

each arm was ascertained, by using the galvanometer, to be equal on the two sides. In the right (unaffected) arm K S Z was produced by a current measuring ten milliampères; in the left only seven and a half milliampères were required for the production of this contraction. In the right arm the closure of a current from twenty cells gave A S O, whilst on the left A S Z was produced with a current from sixteen cells.

From the character of the symptoms, coupled with this evidence of heightened excitability, there can be no doubt of the occurrence of neuritis in this case, and I believe it to be dependent on a gouty condition of the patient's blood. In his occupation, which consists in handling and putting in paper metal goods which have been recently soldered, he has been exposed for forty years to a certain small amount of the influence of lead, which would predispose him to gout, an acute attack of which occurred, as I have said, whilst he was under observation. The disorder of sensibility, by which alongside of some anæsthesia for common sensation there was hyperæsthesia for temperature, is curious, but by no means uncommon. I have referred to it on more than one occasion as occurring in rheumatic neuritis, and have also observed it in hemiplegia with sensory disturbance. It was marked in a case of multiple neuritis, which I shall have to speak about hereafter.

No change whatever may be found in the electrical reactions:

A man, aged fifty, whose father had suffered from gout, and who had himself had three attacks of typical gout, complained of loss of power in the left arm. There was pain in the shoulder and down the left arm, with a slight tendency to puffy swelling of the limb, tingling in the fingers, and later a little herpes in the forearm. The electrical reaction in the nerves appeared unaltered.

I believe that the cases which have thus been briefly alluded to are examples of slight neuritis. In almost every instance they occurred in persons with known gouty antecedents. When we remember the tendency that gout has to cause local inflammation, it seems reasonable to suppose that local irritation from the presence of urate of soda might cause inflammatory action in the trunks of the nerves. One can readily understand, indeed, that when urate of soda is present in the blood it may be liable to find its way into the lymph-spaces which are in immediate connection with the bundles of nerve-fibres, and there set up inflammation. The difficulty is to say why this does not always happen, not to explain its occasional occurrence. Nor is it easy to give a reason for its limitation in such circumstances to one small part of the frame.

I am not alone in feeling this difficulty. In

his recent work on gout, Dr. Roose remarks :
" Sciatica and facial neuralgia sometimes alter-
nate with articular gout. We may assume that
the pain is due to hyperæmia and œdema of the
neurilemma, but why only certain branches of
a nerve should be affected as a result of the
constitutional disorder is a question which cannot be solved."*

It is not without express reason that I have
brought before you such strongly contrasted
cases as one of typical neuritis and others which
are largely dependent upon concomitant circumstances for evidence that they probably belong
to the same class, though of a very different
degree of intensity. When we come to the
variety of neuritis which shows itself in multiple
form, engaging often the peripheral parts of all
the limbs as well as sometimes various cranial
nerves and nerves of the trunk, we shall find
that the symptoms often appear to diverge so
widely from those of typical neuritis as to throw
some doubt upon their relation to that disease.
The various examples to which I have referred
may prove not unimportant steps towards a
recognition of the obscure forms which the symptoms of neuritis may assume.

* ' Gout and its Relations to Diseases of the Liver and
Kidneys,' by Robson Roose, M.D., 1885.

It is not my intention to dwell upon the various forms of localised paralysis, which are well recognised, thanks especially to the graphic teaching of Duchenne. One of the objects, in the brief reference which has been made to certain examples, has been to call attention to conditions which may easily give rise to suspicion of central disease, as indeed had occurred in more than one of those which I have related. This is a point which has been somewhat lost sight of in recent times, although Graves, in his great work on Clinical Medicine, directed especial attention to it. Another reason for occupying time with these narratives has been with the view of showing that in cases of paralysis, in which the lesion certainly occupies the peripheral nerves, we may find a singular diversity of symptoms. As will have been remarked, pain is sometimes present, and sometimes absent, numbness may be slightly or strongly pronounced, muscular atrophy, which is sometimes conspicuous, may be entirely wanting, whilst the results of electrical examination may vary to a remarkable extent. We shall find that the same variety is apt to mark cases in which not one nerve-trunk or plexus alone is the seat of lesion, but where there is a more or less universal affection of the peripheral nerves. The name of progressive multiple neuritis has been given by Leyden to this disease, which, although long since observed, has only been clearly differentiated and referred to its pathological source during the last few years.

It did not escape the attention of Graves, and his reference to the disorder is so graphic and to the purpose that I make no excuse for quoting it :

" One of the most remarkable examples of disease of the nervous system commencing in the extremities, and having no connection with lesions of the brain or spinal marrow, was the curious ' épidémie de Paris,' which occurred in the spring of 1828. Chomel has described this epidemic in the ninth number of the ' Journal Hebdomadaire,' and having witnessed it myself in the months of July and August of the same year I can bear testimony to the ability and accuracy of his description. It began (frequently in persons of good constitution) with sensations of pricking and severe pain in the integuments of the hands and feet, accompanied by so acute a degree of sensibility that the patients could not bear these parts to be touched by the bedclothes. After some time, a few days, or even a few hours, a diminution or even abolition of sensation took place in the affected members ;

they became incapable of distinguishing the shape, texture, or temperature of bodies, the power of motion declined, and finally they were observed to become altogether paralytic. The injury was not confined to the hands and feet alone, but advancing with progressive pace, extended over the whole of both extremities. Persons lay in bed powerless and helpless, and continued in this state for weeks and even months.

   *   *   *   *

" At last, at some period of the disease, motion and sensation gradually returned, and a recovery generally took place, although in some instances the paralysis was very capricious, vanishing and again reappearing.

" The French pathologists, you may be sure, searched anxiously in the nervous centres for the cause of this strange disorder, but could find none; there was no evident lesion, functional or organic, discoverable in the brain, cerebellum, or spinal marrow."

This account, which was published by Graves forty years ago, has been strangely overlooked by most of us.

There is now ample evidence that a more or less widely spread paralysis may depend upon a degeneration of the nerve-fibres themselves, most pronounced towards the periphery and independent of any recognisable change in the nerve-centres or roots.   Such cases may occur in connection with chronic alcoholism, diphtheria, enteric fever, syphilis, tuberculosis, and apparently the widespread disease of Japan, called Kakké or Beriberi, is an endemic form of the same affection.   There is some reason, too, to think that exposure to cold, which is a frequent antecedent, may be a factor in the production of the disease.   Occasionally also cases occur in which no etiological cause whatever can be traced.   In this class of neuritis we have not to deal with gross changes in the nerve-trunk. The alterations are in great measure confined to the nerve-fibre itself, and are usually only recognisable under the microscope.

I propose in the next lecture to give an illustrative sketch of the symptoms and course of multiple degenerative neuritis, and discuss afterwards some of the modifications which are apt to occur in one or other form of the disease.

LECTURE II.

Multiple neuritis—Two cases with syphilitic history—
Grainger Stewart's cases—Morbid anatomy—Sym-
ptoms of multiple neuritis—Comparison with Kakké
—Alcoholic paralysis—Examples—Ataxic symptoms
—Pains and hyperæsthesia—Degenerative changes in
peripheral nerves the immediate cause of paralysis—
Lancereaux's investigations—Pains may be absent—
State of the knee-phenomenon—Nystagmus—Dr.
Churton's case—The peripheral neuro-tabes of Déje-
rine probably of alcoholic origin—In alcoholic paralysis
lower extremities most but not exclusively affected—
The symptom "dropped feet" should always suggest
inquiry as to habits of intemperance—So "wrist-
drop" suggests inquiry into possible exposure to lead
—Not in alcoholic paralysis alone is found tendency
to preponderating affection of the lower extremities—
Seen in multiple neuritis from other causes—Also in
Kakké—A paraplegic form of neuritis from unknown
cause—Examples.

In 1874 I brought before the Clinical Society
a case which I have since had reason to think
must have belonged to the class of multiple
neuritis. I recognised at the time that the
symptoms must, from their anatomical distribu-

tion, depend upon altered conduction in nerve-fibres, and not upon central changes, but was disposed to refer these alterations of conductivity to pressure upon the roots of nerves by inflammatory changes in the membranes. At that time the idea that paralysis of a more or less universal kind could depend upon lesion of the periphery of the nerve-trunks, the central portions being undamaged, was not entertained. The case is so important that I think it well to give a brief account of it, as well as of another, manifestly of similar character, which came before me in 1879. Dr. Ross, in referring to these cases in the second edition of his admirable work on the ' Diseases of the Nervous System,' has also expressed the opinion that they were examples of progressive multiple neuritis. I have described them* under the head of " rapid and almost universal paralysis," remarking that " it was convenient to use some such general title for the designation of certain cases which are of extreme interest and importance, but the true pathology of which we have yet to learn." Since the first example, which was published in the ' Transactions of the Clinical Society ' in 1874, I have met with

* ' Clinical Lectures on Diseases of the Nervous System,' p. 301.

several, some of which are referred to in my book. The cases that I am about to quote in brief abstract are the most typical that I have seen, and the most useful, therefore, for illus‧trating the subject.

W. H—, a working man, aged forty-four, of previous good health, was brought to me at the hospital in January, 1873, in the following condition. He had double facial paralysis, total absence of power of voluntary contraction in the muscles of either leg, the grasp of both hands almost entirely lost, and partial paralysis of respiration and deglutition. There was incomplete paralysis of the right external rectus muscle and of the soft palate, especially on the left side. There was but little movement of the diaphragm, and the intercostal muscles were likewise acting so imperfectly that the patient could not lie down in bed. His sterno-mastoid and trapezii muscles acted freely. Cutaneous anæsthesia was more or less general throughout the trunk, extremities, and face—the tips of the fingers being especially numbed. The plantar reflex was absent in each foot. There was slight power of voluntarily contracting the muscles on the front of each thigh, but he was unable to contract in the least those on the front of either leg below the knee. A sense of numb‧ness and weight was complained of in each leg, and occa‧sionally a "throbbing ran down the left thigh and calf." For the first two or three weeks also he had suffered from "pins and needles" in his legs. But at no time appa‧rently had there been any actual pains in his extremities or involuntary muscular contraction. The power of the sphincter ani was normal, that of the bladder impaired to a slight extent. The muscles about the mouth showed the reaction of degeneration. In those of the arms the reaction to faradism was greatly diminished, whilst in those of the legs, below the knees, it was quite absent; in the left thigh it was greatly diminished. (The right lower extremity was lame and wasted from an old attack of infantile paralysis.) But in no part of the upper or lower extremities was there increased action to slow intermissions of the galvanic current. In the face, however, this was marked. The facial muscles reacted to interruptions of a current from six cells (Stöhrer). His attack had commenced one month previously with numbness in the finger-ends, followed on the same day by weakness in the legs, which increased next day and was then accompanied by numbness about the calves, thighs, and buttocks. The weakness increased day by day, and a week after the beginning of his illness he had the sensation of a tight band round his abdomen. A few days later he could use neither arms nor legs. The difficulty of swallowing was not observed till a fortnight after the onset. There had been no fever. There was nothing abnormal in the mental condition, nor in the heart, lungs, and kidneys. The patient was at once admitted and placed on a water-bed. For twenty-four hours his condition was one of imminent danger from the state of respiration. As there was a syphilitic history, he was treated with iodide of potassium, and later with mercury. He soon began to improve, and in six months was able to resume his employment. A few months later I showed him at the Clinical Society, entirely recovered.

On April 14th, 1879, T. O—, aged forty-four, was admitted into hospital with paralysis of all four extremities and both sides of the face, together with inability to swallow solids. The respiration was mainly upper thoracic, and there was some loss of control over the bladder and sphincter ani. His grasp was feeble; he was unable to stand; as he lay in bed he could move one foot across the other, but could not lift either more than three or four inches off the bed. There was great muscular flaccidity. He did not know where his legs were. The knee phenomenon was absent on each side. Over the right side of the face there was complete loss of sensibility to touch and pain, with apparently increased (but at all events well-retained) sensibility to heat and cold. The anæsthesia was likewise observed, though to a less extent, over the left side of the face, and also, though here again in a less complete degree, on his forehead (Fig. 1). He complained of great pain in the right half of the forehead, spreading towards the vertex. Below the middle of the

Fig. 1.

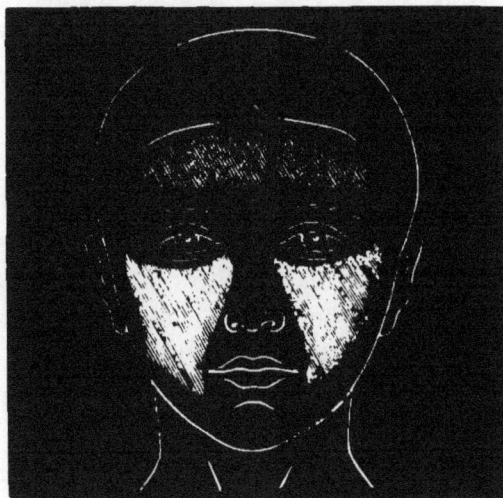

forearm on each side there was almost entire loss of sensibility to touch and pain, whilst heat and cold were well recognised. Where the alteration in sensibility began there was what the patient described as a "band-like feeling around the arm." In the tips of his fingers there was a constant tingling sensation, and anything which he touched with them felt hot (Fig. 2). In his lower extre-

Fig. 2.

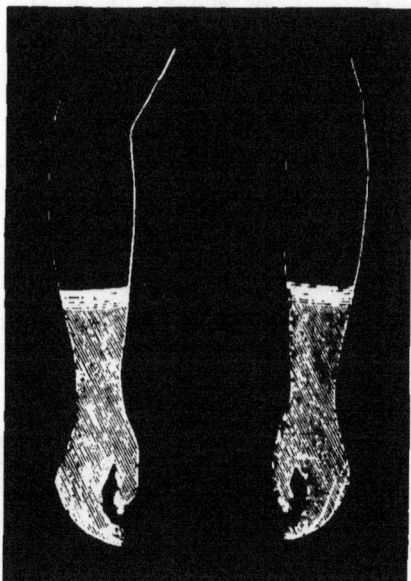

mities sensibility was also greatly modified. Below the middle of the thigh on each side neither a touch nor the prick of a pin could be recognised. Over the whole of both feet, as well as half-way up the legs, and on the posterior surface of the rest of each lower extremity, sensibility to heat and cold appeared intensified. Over the whole of the trunk, and in the extremities down to the boundaries described, cutaneous sensibility was normal. He had frequent pains like knife-stabs in the lower extremities, and the legs would twitch when they occurred. At other times the pains were of a dull heavy character

(Fig. 3).* In this case facial paralysis on the left side seems to have been the first symptom, which was observed by his friends a month before admission. The patient

FIG. 3.

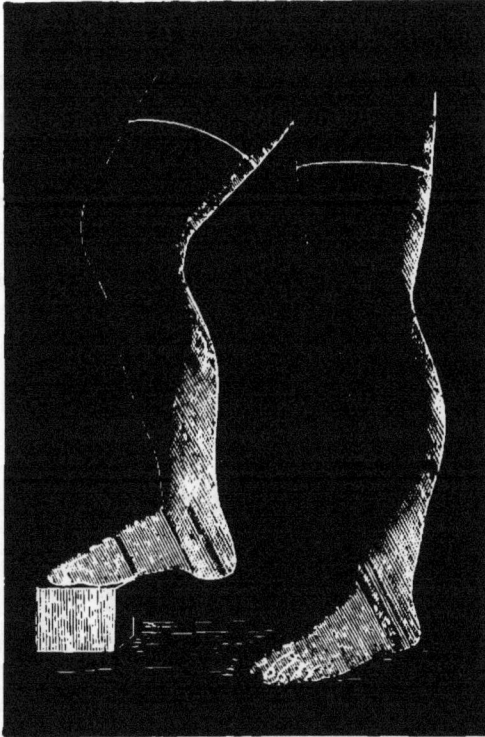

knew nothing of this, and felt quite well till a fortnight before I saw him. He then noticed "pins and needles" in his hands and feet. Three days later he had diplopia, and his legs became weak. In five more days he could not

* The illustrations are from diagrams made at the time by Mr. A. E. Broster, then resident medical officer, and exhibited at the Clinical Society, before which the patient appeared when recovered.

walk, and there was difficulty in swallowing. In another three days he could not dress himself. On admission, and for a few days afterwards, reaction to induced currents was almost entirely absent in the muscles of the face, and also in the thenar eminence and interossei of each hand; it was lessened, though not to the same extent, in the muscles of the front and back of the forearms. There was very slight reaction to induced currents in all the muscles of the lower extremities. Under active mercurial treatment the patient entirely recovered in about six months. On August 10th the knee-phenomenon was found to have returned in the right leg, and three days later in the left. This patient was likewise shown at the Clinical Society.*

There is no doubt that the patient, W. H—, was a temperate man. As regards T. O—, he is described in my notes as having "lived fast and drunk fairly," whatever that may mean. Both had had syphilis, and recovery in each case was absolutely complete under specific treatment. Diphtheria could be entirely excluded. I would draw attention to the fact that in the first case the patient suffered from girdle pain, a symptom which is usually considered as pointing distinctly to disease of the spinal cord or its membranes. In the second case, again, what was equivalent to girdle pain was felt in the two forearms, at the point where the anæsthesia ceased. It is a curious circumstance,

* 'Transactions of the Clinical Society,' vol. xiii.

and well illustrates the difficulty there is in quitting well-worn grooves of opinion, that, although I recognised that the lesion must be one of nerves, I failed to see in the diagrams of the anæsthesia powerful evidence that the affection must involve the periphery of the nerves, and not their roots, nor did any member present at the meeting at which they were shown suggest such an explanation.

In March, 1881, Dr. Grainger Stewart brought before the Medico-Chirurgical Society of Edinburgh three cases of paralysis of the hands and feet from disease of the nerves.*

The first was that of a clerk, aged fifty-two, who got bruised and shaken from a fall, and on the following day had some " rheumatic " pains which came and went during the succeeding week. A fortnight later there was tingling and numbness in the hands and feet, with stiffness in the joints of the hands, and difficulty in grasping and walking, swelling of the feet and hands, with severe cutting pains in them. All these symptoms gradually increased until admission. When examined his temperature was under 100°. There was severe cutting pain of an intermittent character in the hands and feet, increased on pressure, especially of palmar or plantar surfaces, and aggravated by movements of the body. With the pain there was numb-

---

* The account which follows has been much abbreviated from the report in the 'Edin. Med. Journal,' April, 1881.

4

ness, but the tingling was gone. Sensibility to touch was much lessened in both hands, slightly in the forearms and was normal above the elbows. It was also greatly diminished and delayed in both feet, and, to a less extent, in the legs. Sensibility to heat and tickling were correspondingly affected. The perception of painful impressions was delayed, but felt acutely, a distinct interval elapsing between pinching the hand and its perception, but a slight pinch was felt to be very painful, and the pain persisted longer than normal and set up considerable reflex movements. The muscular sense in the hands and feet was much diminished.

Voluntary motion was greatly impaired in the hands and feet, all the groups of muscles being weakened, but those of the fingers and toes completely paralysed. There was diminution of electric excitability in proportion to the degree of paralysis.

There was glossiness of the skin over the backs of the fingers and some appearance as of bruises over the toes, but no œdema. The muscles were diminished in volume. There was acute pain in passive movements of the hands and wrists. The case was treated at first by ergot and then by the induced current. A degree of improvement soon manifested itself, and by the end of September the patient could walk about, and the muscles of the extremities were recovering their volume.

In another case the patient's illness began with chilliness, numbness, and weakness in both legs; next day, "pins and needles" and increasing weakness. Some days after the feet and hands became similarly affected. Pain, though not of severe character, was felt in the small of the back, but no mention is made of pains in the hands and feet, as in the last case. Sensibility to touch was distinctly impaired in the parts which suffered from coldness, formication, and numbness, *i. e.* in the feet and tips of the fingers.

Voluntary motion was lost in the toes, very imperfect at the ankles, normal at the knees and hips. The hands, especially the right, were much weakened, and the movements of the wrists, fingers, and thumbs, although feeble, were not lost. The patient could neither walk nor stand unsupported. At one time it was necessary to draw off the urine by catheter. The patellar tendon-reflex was absent on each side.

After four months this patient was likewise recovering.

The third patient was thirty-one years old. In him a weakness of the legs appears to have been the first symptom noted, which was followed by " prickling " in the legs, which increased, and in a short time was accompanied by a similar feeling in the fingers and hands also, with loss of power and stiffness.

This man likewise had pain of a tingling character in both legs from the knee to the dorsum of the foot, but no girdle pain or formication. Sensibility to touch was diminished in the legs below the knees and in the hands. There was delay also in the transmission of impressions. Patellar tendon-reflex was absent. Voluntary motion was greatly impaired in the legs and hands. Electric excitability was much diminished in the legs and forearms, especially in the extensor muscles. Any attempt to use the muscles caused great pain.

The cerebral and mental functions of this patient were somewhat impaired. He was drowsy and his memory was imperfect. He was improving when an attack of croupous pneumonia carried him off. No mention is made of his habits. Was it not an example of alcoholic neuritis?

In this case microscopical examination showed very notable changes in the median, ulnar, and tibial nerves.

The axis-cylinders had undergone degeneration. On the other hand, two cords of the brachial plexus which were examined showed no evidence of such changes as were found lower down in the nerves of the limb. Most of the bundles of nerve-fibres appeared normal. There were no such alterations of the axis-cylinder as were found in the nerves of the forearm. Certain changes were found in the cord, but these apparently were of a secondary character. They occupied the columns of Goll, and the most superficial and posterior part of the lateral columns, and this only in the cervical region and to a less extent in the lumbar enlargement. Neither the grey matter nor the nerve-roots were involved.

" In the nerves," Dr. Stewart remarks, " the lesion was extremely distinct. It involved a large proportion of the nerve-fibres of the affected portions of nerves, and consisted in a breaking up of the axis-cylinder. To the naked eye the nerve appeared quite normal, the affected parts of the median and radial being undistinguishable from the unaffected parts further up."

Leyden describes an acute or subacute inflammatory process as affecting the nerve-trunks, especially the radial and peroneal, in a case of this description. The nerves appeared in some cases swollen, reddened, and hyperæmic, but sometimes also unaltered to the eye. The nerves themselves presented signs of sclerotic atrophy. The sheaths were thickened, the primitive nerve bundles contained exceptionally few medullated fibres, most of them atrophic and composed of thickened axis-cylinders. About the blood-vessels in the fatty tissue between the nerve bundles, and also in these themselves, there lay numerous heaps of yellow granular pigment; evidence of a previously existing bloody infiltration.

This degeneration of nerves extended downwards into the smallest twigs of muscular nerves. The muscles themselves showed a moderate degree of myositic atrophy. Upwards this disease ceased about the middle of the upper arm. The anterior roots were quite intact; the cords also intact, and the great ganglion cells well preserved. No disease was to be found in the nerves of the lower extremities.

The symptoms had been fever, sharp pains, involving the upper and lower extremities. Pains extended from the elbow and knee-joints to the fingers and toes, either spontaneously or as a result of pressure.

There was loss of power below the knees and in the forearms, with considerable atrophy of muscle in the forearms, especially in the extensors, but trifling in the leg and foot. In the lower extremities, the affection after some months was completely recovered from. But it lasted in the upper; the atrophy increased and the difficulty of using the limbs became greater. The muscles showed reaction of degeneration.

The patient died with disease of the kidneys one year after the commencement of the illness.

In another case the nerves showed fatty degeneration and atrophy. A great part of the fibres had their medullary sheath entirely lost, and the nerve was changed into band-like axis-cylinders which showed varicose swellings in different parts.

It will not be necessary for me to multiply illustrative cases of the kind, but I may say that many such have lately been recorded both in Germany and France. The two cases of my own show a considerably greater diffusion of the lesion than was observed in those reported by Dr. Grainger Stewart, but in other respects they are strictly comparable, and there can be no doubt that a lesion of peripheral nerves similar to that which the fatal termination in one of Dr. Stewart's cases enabled him to disclose was also the cause of the symptoms in my patients, who perfectly recovered. The double facial paralysis in my cases, whilst apparently complicating the diagnosis, really aids it. The facial muscles showed well-marked " reaction of degeneration," evidencing lesion either of the facial nucleus or of the trunk of the portio dura. The simultaneousness of the paralytic phenomena in other parts of the body make it evident that they all (facial included) depended on strictly similar lesions. Therefore, had lesion of the facial nucleus been in question, the motor symptoms in the extremities would have been due to lesion of the ganglion cells in the anterior horns of the cord, homologous with the facial nucleus in the bulb. But the sensory symptoms were as strongly marked as the motor, and lesion of the anterior ganglion cells could not explain these. We are obliged, therefore, to seek the cause in the nerve-trunks; and the strict limitation of the anæsthesia, as shown in the diagrams, coupled with the corresponding situation of the most marked paralysis, is conclusive evidence that the lesion was in the periphery of the nerves.

The fact that in many examples of multiple neuritis the spinal cord and the roots of nerves are perfectly free from pathological change, which is confined to a greater or less extent of the periphery of the nerves, is very startling and difficult of explanation. One cannot help feeling that notwithstanding the absence of observable lesion there must be some change in the central nervous system to influence a coincident affection of the periphery of so many nerves. An ingenious suggestion has been made by De Watteville in reference to certain cases of lead paralysis where this limitation of morbid change to the peripheral nerves sometimes occurs. He inquires whether, in these circumstances, there is not a *functional* lesion of the cord which interferes with the trophic influence which ought normally to be exerted by it upon the motor nerve-fibres, and consequent degeneration. More recently Erb has suggested the same explanation for the paradoxical condition obtaining in cases of multiple neuritis generally.

But one would not expect, if this were the case, that the lesion of nerves would continue as it sometimes does after long illness strictly confined to the periphery. Were the functional activity of the cord to be long suspended the atrophy might well be looked for throughout the length of the motor fibres. For my part I should rather seek to explain the fact by the hypothesis that a toxic influence of some kind is exerted upon vaso-motor centres in the bulb and cord. The effect of this would be to give rise to some alteration in the calibre of minute arteries—those in which the muscular element is relatively most developed. These belong to the periphery of the arterial system, and in general terms correspond with the periphery of the distribution of nerves. It is conceivable, therefore, that the supply of blood to the periphery of nerves might in this way be diminished by an irritative influence exerted upon vaso-motor centres in the bulb and cord. Notable diminution of blood supply sufficiently long continued would occasion degeneration of the essential element of nerve-fibre. I merely throw this out as a suggestion. It is impossible to discuss the point just now with advantage.

SYMPTOMS OF MULTIPLE NEURITIS.

In cases of acute or subacute multiple neuritis, it very commonly happens that the first symptom noted by the patient is a feeling of " pins and needles " or numbness in the feet, and about the same time, or a little later, in the finger ends.　In others there is less acuteness in the attack, and vague pains of a " rheumatic " character have been complained of before the occurrence of numbness.　Fever is not usually a marked symptom, but occasionally there is considerable elevation of temperature. It is difficult to speak very definitely on this point, because from the insidious mode of onset which often characterises even the acute cases, the patient is not usually examined with the thermometer until some days after the commencement of his illness.　More often than not the patient in the early stage of the disorder tries to go about his usual avocations, but finds day by day an increasing difficulty in doing so.　The numbness and deadness which had commenced in his feet and fingers gradually spread up the extremity, his legs appear to grow heavy so that he cannot move them quickly, his arms become more and more powerless, and in a few days he cannot stand or help himself in any way. The disease tends to affect both sides of the body symmetrically, though sometimes there are considerable differences in the severity of the symptoms as displayed on either side. In severe cases, we find not only the muscles of the extremities, but also those of the trunk, becoming more or less powerless; there may be facial paralysis, and some of the muscles of the eye may become involved; swallowing and respiration may also become affected; and death may occur, with signs of the vagus becoming implicated.

But it much more commonly happens that at a certain point, which differs remarkably in various cases, the climax of the attack is reached, and then each day brings with it signs of amelioration, until in many cases absolute recovery takes place. But here, again, it is difficult to give a general sketch which shall be consistent with the many varieties which occur. The amelioration in some instances may take place with such quickness as to make it appear doubtful whether any serious organic lesion could have been present. In others, the amount of improvement which each day brings is so slight that the prognosis remains for a long time doubtful. Or there may be condi-

tions between these two extremes. In that
stage of the disease in which the patient is able
to walk about, the gait is apt to be ataxic.
The duration of the illness may be from a few
weeks to many months, or even, if the sequelæ
be reckoned, some years. During the entire
illness there is in the large majority of cases
great flaccidity of the paralysed muscular
system. There is no doubt that this is the
rule. But I have seen several exceptions, and
to these I shall have to refer more particularly
later on. In many cases there is distinct muscular atrophy. This is especially marked in
the muscles of the leg below the knee, and in
those of the hand and forearm. As the patient
lies in bed, even from a very early stage it is
characteristic of this disease that the feet are
"dropped," so to speak, the power of dorsal
flexion of the foot being the first to disappear.
And so also with the upper extremities. The
wrists are "dropped" exactly as is seen in
cases of lead palsy. The flaccidity of the muscular tissue, its tendency to atrophy, and its
behaviour to electric stimuli, likewise cause a
strong resemblance to this well-known form of
toxic paralysis—a resemblance due to the circumstance that a similar condition of the peripheral nerves may be due to the influence of

lead. When the facial muscles are paralysed, the cheeks fall in bags, the food collects in them, and there is absolute want of power of expression. The soft palate, when it is involved, hangs loose, and is unable to be lifted, so that fluids regurgitate through the nostrils, and the voice has a nasal character. In very slight cases the electrical excitability of the muscles may be unchanged; and in severe examples, if the muscles about the body be severally tested, you will find a great variety in their response. In some, faradic excitability will be slightly, in others greatly, lessened. Along with great lowering of faradic excitability in the first of my cases there was no increase of excitability by the galvanic current. Löwenfeld, on the other hand, has observed isolated lowering of galvanic excitability, the response to induced currents remaining normal.

In others, again, " reaction of degeneration " will be distinctly marked. This is especially likely to be the case as regards the intrinsic muscles of the hands, and the anterior muscles of the leg below the knee. The patellar tendon reflex is almost always lost. It is usual to find the cutaneous reflexes of the sole, abdomen, and cremaster more or less weakened, or altogether absent. As a general rule the functions of the bladder and rectum are not disordered, but in severe cases there is a loss of control over the sphincter ani, and if the patient is not quick to answer to the call of his bladder the urine runs from him. Or there may be some delay in passing urine when the desire to do so is present.

It has been suggested by Leyden that in the exceptional cases in which bedsores and paralysis of the sphincters occur we have to deal with an affection which is not limited to the peripheral nerves, but extends also to the spinal cord. Occasionally more or less œdema of the extremities is to be observed, and trophic changes in the skin have been noted, especially a glossiness like that described by Paget as the result of injuries to the trunks of nerves.

On the sensory side we may expect to find pains which are often of lightning character, coming and going in sudden darts like stabs of a knife, and recalling those which are characteristic of tabes dorsalis. Or they may be described as " gnawing " or " burning " or " like molten lead in the veins." They are usually more pronounced in the lower than in the upper extremities. It is commonly found that great tenderness of the muscles is complained of when these are grasped by the hand. The patient himself will sometimes describe a sensation of aching in the muscles, and very commonly, indeed, a feeling of " numbness," " deadness," or " pins and needles," which are referred especially to the hands and feet.

Leyden calls attention to the fact that sensations of " pins and needles " or " deadness " are apt to be produced by pressure on the nerve-trunks, and remarks upon the interesting bearing of this circumstance upon the pathology of the affection. Remarkable differences may be found as regards the affection of various modes of sensibility of the skin. There may be, as we have seen, in the lower half of the forearm and hand, entire loss of sensibility to touch and pain, whilst heat and cold continue to be well recognised, or even exaggerated in intensity. Or we may find exquisite hyperæsthesia, so that not only is a touch unbearable, but even a current of air excites the greatest torture. In other cases there is only a "muffling" of common sensation, and in some no disorder of sensibility is to be noted. There is almost always a striking absence of tendency to bedsores. The mental faculties may be expected to remain entirely free from disorder. It will happen, however, in those cases which are connected with alcoholism that intercurrent affections of the brain or its membranes may produce their peculiar effects. These complications are not unlikely, unless remembered and allowed for, to cause some obscurity in the diagnosis.

I have said that many of these cases recover perfectly. On the other hand, the process of recovery may remain permanently incomplete owing to irreparable atrophy of muscular tissue having taken place. It is in cases marked by a great amount of muscular atrophy that we are apt to meet with apparent exceptions to the rule that the paralysis is of a flaccid character. From the circumstance that all the nerve-fibres supplied to a limb are not affected with equal severity, the amount of atrophy in different sets of muscles will necessarily differ. Some will lose but little tissue or none at all; others, their antagonists, may be profoundly affected. Duchenne has shown how, as a consequence of loss of power in a certain muscle, the limb, obeying the tonic power of its antagonists, whose exaggerated movement is no longer held in check, is ultimately dragged in the direction of these latter. Hence occur faulty positions and deformities. The contractures are rigid and not to be overcome by passive movement. So you will often see the foot stiffly pointed from unopposed contracture of the sural muscles due to atrophy of those lying on the front of the leg, and analogous rigid malpositions in the upper extremities.

Attention has been called to the strong resemblance to multiple neuritis in its clinical features which is borne by the disease called Kakké or Beriberi.

Hirsch * reports that after prodromal malaise there is a gradually increasing paralysis of the lower limbs which extends also to the upper extremities, and the patient is more or less completely disabled by loss of power. On the sensory side there are at first paræsthetic symptoms, tickling, creeping feeling, or " pins and needles " and burning sensation, particularly in the feet and legs, muscular tenderness (especially in the muscles of the calf). Later on, the sensory disorder takes the form of anæsthesia, and is especially distinguished by loss of the sense of pressure, and of heat and cold. These symptoms, like those on the motor side, are noted earliest and most uniformly in the lower limbs, extending afterwards in a smaller proportion of cases to the upper extremities. There are also disorders of the blood-making organs and circulation. The patients wear an anæmic

* 'Handbook of Geographical and Historical Pathology,' New Sydenham Society, ii, p. 571.

look and complain of palpitation and dyspnœa. There are cardiac murmurs with increase of the area of cardiac dulness (from dropsy of the pericardium or dilatation of the heart); the pulse is small and very compressible. The urine is always diminished but never contains albumen. Such are the symptoms which characterize what is called the paralytic or dry form of Beriberi, but in another class more or less dropsy is added to them. The œdema usually shows itself first in the ankles and legs, extending gradually until it at length becomes general anasarca; then follow dropsical effusions into the serous cavities, always into the pericardium, more occasionally into the pleura or peritoneum, and rarely into the cerebral and spinal arachnoid. It would appear that whilst that form which is characterized by the presence of dropsy is rapidly fatal, the other, called the "atrophic" form, but rarely causes death.

Dr. Theobald Palm,* who was for ten years resident in Japan, says, " I soon became aware of a large number of patients who came to me with symptoms of defective innervation of the legs. They complained of numbness in the

* Kakké, 'Edin. Clinical and Pathological Journal,' September, 1884.

lower extremities, which they generally described
as a feeling as if a piece of thin tissue paper
were spread over the skin; of slight loss of
power, showing itself by inability to walk any
distance without inordinate fatigue, a tendency
to stumble and for the knees to give way.
They experienced a difficulty, especially in
going upstairs, and sometimes in holding the
thong of the wooden clog usually worn by the
Japanese, which passes between the great toe
and the next.   Some patients dropped the point
of the foot in walking, showing a paralysis or
paresis of the flexors of the foot and extensors
of the toes.   When the foot was planted evenly
on the floor they had little or no power to raise
the toes from the ground, or if they could raise
the toes little force was required to press them
down.   They had also a trace of œdema over
the tibiæ or about the ankles.   In many cases
there was tenderness of the muscles of the calf,
which were in some instances hard and swollen,
in other cases abnormally flabby and apparently
partially atrophied.   In almost all of these
patients there was an absence or marked diminution of tendon reflex at the knee.   Beyond
the above symptoms they seemed to be in
average good health.   Some of them complained of vague dull pains in the legs."

Scheube* mentions that in the upper extremities, as well as in the lower, it is the extensor muscles which are most affected. The feet are "dropped" by their weight, and it is only rarely that a position of varo-equinus is caused by contraction of the unopposed sural muscles. He notes the decrease or loss of faradic excitability and disorders of sensibility closely resembling those I have described, and remarks that it is especially the tactile sensibility and the muscular sense which are affected. In advanced cases there is pain on pressing masses of muscle. In about a third of the patients whom he examined the knee-jerk was absent. Frequently it disappeared several days before the apparent onset of the disease, and did not return till some months after recovery from all other symptoms. He did not note any laryngeal symptoms, but is disposed to refer the quickening of the pulse and the vomiting which may be frequent towards the end to neuritis of the pneumogastric. Recovery occurs in the majority of cases, but in a certain number this is not complete.

Dr. Harada reports details of autopsy in two cases of Kakké in its subacute form. Amongst other points he noted degenerative changes in

* 'Die Japanische Kakké (Beriberi),' Leipsig 1882.

the muscles. The nerves of the lower extremities showed overgrowth of connective tissue in the epi- and endo-neurium. The vessels of the nerve-trunks were thick-walled, with narrowing of the lumen, or hyperæmic with widening. The nerve-fibres were very irregular, being either sodden and often granulated, or atrophic; the axis-cylinder was often very thin. The changes were very considerable in the peripheral part of the tibial and peroneal nerves, less in the sciatic and crural; they were also distinct in the renal nerve and vagus. The brain was normal. The substance of the spinal cord was indistinctly seen under the microscope, owing to the presence of fibrinous exudation. Some of the ganglion cells in the anterior horns were atrophic, round or irregular in form, granular, often without nucleus, the processes sometimes strikingly thin. Some of the vessels showed thickening of the intima with round cells.

So close is the resemblance between this endemic disease and the " spontaneous " form of multiple neuritis met with in Europe, that R. H. Pierson, of Dresden, is inclined to consider the latter as sporadic cases of the Oriental disease. Whilst on this point I would remind you that in anæsthetic leprosy we have a train of symptoms pointing to lesion of peripheral nerves, loss of power, muscular atrophy, trophic changes in the skin and deep tissues, with cutaneous anæsthesia. In ten such cases Vandyke Carter found the brain, spinal cord, and roots of the nerves healthy, whilst the nerve-trunks were swollen, dull red, or grey, or semi-translucent, rounded, and firm. According to Virchow, the morbid process begins with a peri-neuritis causing proliferation of cells and consequent atrophy of the essential nerve-elements by pressure. In a patient affected with anæsthetic leprosy, whose case I brought before the Clinical Society in 1870, and who died an accidental death, the nerve-trunks of the affected limbs were found swollen, and microscopic examination disclosed great over-growth of cells in the connective tissue, with atrophy of nerve-fibres. I may remark here, by the way, that I long ago suggested to the late Dr. Anstie that the scattered cases of localised muscular atrophy met with in this country might possibly be the lineal descendants of the leprosy which once abounded in Great Britain. But I must not be tempted to travel further in a by-way full of attraction for the exercise of the scientific imagination.

Let us pass on now to another form of progressive multiple neuritis—that which is apt to take place in alcoholic patients.   The earliest reference to this disease that I have met with occurs in a short essay, ' History of Some of the Effects of Hard Drinking,' by the well-known Dr. Lettsom, which was published in 1789.*   In describing the symptoms which are liable to arise from an acquired habit of spirit drinking he writes, " The lower extremities grow more and more emaciated, the legs become as smooth as polished ivory, and the soles of the feet even glassy and shining, and at the same time so tender that the weight of the finger excites shrieks and moaning.   .   .   . The legs and the whole lower extremities lose all power of action; wherever they are placed there they remain till moved again by the attendant; the arms and hands acquire the same palsied state, and the patients are rendered incapable of feeding themselves.   .   .   . They talk freely in the intervals of mitigation, but of things that do not exist; they describe the presence of their friends as if they saw realities, and reason tolerably clear upon false premises."

Some thirty years later Dr. James Jackson,

---

* I am indebted to my friend Dr. Thomas Morton, late President of the Harveian Society, for the opportunity of consulting this work.

of Boston, published an account of the disorder,* and Magnus Hüss, of Stockholm, drew attention to it in 1852. Since then important contributions to our knowledge of the subject have been made by Lancereaux and Leudet in France, and by Reginald Thompson, Handfield Jones, and Wilks in England. But little has appeared on the subject in England since Dr. Wilks' communication to the ' Lancet ' in 1872, except a paper at the Royal Medical and Chirurgical Society by Dr. Broadbent in the course of last year, and more recently valuable contributions from Dr. Dreschfeld and Dr. Hadden. My personal observation of the disease dates from a case which I attended in 1870 in consultation.

The patient was a lady who had for years been addicted to great alcoholic excesses, consuming large quantities of brandy. When called to her she was in bed suffering from considerable loss of power in the upper and lower extremities. The hands were dropped at the wrists, and she could not extend them. The feet, too, were likewise in a dropped condition, and there was no power of dorsal flexion. There was much mental disturbance, and such loss of memory that the patient could give no intelligible account of the duration of her illness. She could move her arms and raise the knees, though with difficulty. The

* Quoted by Dr. Dreschfeld, 'Brain,' part 32.

functions of the bladder and rectum were not interfered
with.   Her naked feet projected from the foot of the bedclothes, and she would not allow anything to be laid
upon them, so exquisite was the tenderness of the skin.
Her constant complaint was of the agonising pains in the
legs, " as though the veins were filled with molten lead,"
and she appealed piteously to those around her for relief
from this suffering.   She was placed in charge of two
nurses, who gave her no more than the very small amount
of stimulant which was allowed, and under careful feeding
and treatment her pains got rapidly less.   There was much
muscular atrophy of the hands and forearms and the anterior tibial muscles, with what is now known as " reaction
of degeneration."   Her hands assumed the typical form of
the " *main en griffe* of Duchenne."   In the course of a
little more than a year she had recovered so far as to be
able to go about by herself, and the movements of the
hands also were practically restored.   The first use which
she made of her liberty was to visit a succession of taverns,
and inaugurate a debauch, which terminated her existence
in the course of a week or two.

This case made a great impression upon me,
for it was at that time a novel experience.
Since then I have seen many examples of varying gravity and with certain differences in the
symptoms.
The following are short notes of a case which
I saw in consultation in September, 1875 :

A female patient, aged thirty-five, lay in bed with her
knees up, and said she could not straighten her legs, but
being partially assisted and by great effort she contrived to

get her legs down nearly to extension, complaining, however, of the pain thereby caused in her hamstrings, which were very taut.

There was partial anæsthesia of the feet and legs up to the knees. Tickling the sole was felt as touch, not as tickling. Her hands, too, were similarly numbed, the fingers looked white, puffy, soddened, and there was tremulousness of the tendons.

She herself dated her illness from a miscarriage two years previously, after which she began to feel curious sensations in the feet as though walking on sponges, with occasional pricking, tingling, and pins and needles. These feelings continued and were later accompanied by pain of a very excruciating character, which began in the toes, especially the big toe, and ran up the foot and leg to the knee, She described the toes as "drawing up" and the pain "as though the skin were being dragged off her flesh with a knife." A fortnight before I saw her she had walked about fifty yards supported on either side. She had been obliged to look continually at her feet or would have fallen.

There was no spinal tenderness on touch or percussion, nor marked stiffness of the spinal column on rising or reclining in bed. She suffered from insomnia, "seeing things about the room," and there had long been retching in the morning. She had no albumen in the urine. Her habits had long been very intemperate.

The necessity of looking at her feet in walking described by this patient recalls a similar symptom so common in patients suffering from tabes dorsalis, and is evidence of ataxy. It is a symptom which has caused not a few cases

of alcoholic paralysis to be mistaken for locomotor ataxy.

In the case of a gentleman, aged forty-three, whom I saw
in 1876, there was paralysis of the bladder as well as of
the lower limbs.   He had been drinking spirits very hard
indeed for three or four years, and previously to that had
always lived freely.   He was paraplegic and complained of
horrible pains, "like something sharp thrust into his legs
and making him twinge again."   The legs did not jerk
nor did he complain of cramp.   There was no pain in the
spine or tenderness on pressure upon it.   His conjunctivæ
were yellow, his hands tremulous.   There was no alteration in the fundus oculi.   The urine was albuminous.
It appeared that seven or eight years previously he had
suffered from a similar attack, from which he had completely recovered.

By my direction, the patient (whom I only saw on one
occasion) was kept without alcohol, and for a week he
appeared, it was said, to be doing well.   His pains became
much less and he slept, but I learned in the sequel that
he contrived to get drink again and shortly afterwards
died.

In 1882 I saw in consultation a lady, of about thirty-
five years of age, who had gradually lost power in the
lower extremities during the preceding six months.   The
muscles of her legs were excessively wasted and powerless;
the feet were dropped and dorsal flexion of the foot was
impossible.   The knee-jerk was absent on each side.
Examination with the induced current showed entire loss
of faradic contractility.   There was also atrophy of the
interossei of the hands.   She complained of pains like

"molten lead" in the feet. In this case there was a history of confirmed alcoholism. Abstinence was enforced, and under appropriate treatment the patient, as I learned some months afterwards, quite recovered.

Some years ago a female patient of middle age came under my observation. She had suffered much from neuralgia in the head, which had been relieved by stimulants. In the preceding summer she had been subjected to great moral and physical strain, and after a few weeks it was noticed that her feet began to swell and she became very weak. (There was no cardiac or renal disease.) Severe vomiting and diarrhœa began, which brought her down very low. Then loss of memory came on and delirium—or rather a vacuity of mind and alteration of manner. About this time she began to suffer from severe paroxysms of pain in her joints and limbs. Menstruation ceased for four months. Some of her joints became contracted, but the general health improved under treatment, and when I saw her the wasted and contracted state of the limbs was the chief ailment to be observed. I found the little finger of one hand contracted to a right angle, the ring finger also, though less contracted, and there was a thickening to be noted about the tendons in the palm of the hand. The extensor communis digitorum responded to the induced current. She could lift the right knee fairly well, but could not straighten the leg owing to contraction of the hamstrings. There was a good deal of wasting of the quadriceps extensor muscle. The vastus internus muscle was only excitable by a strong induced current. The foot was in a position of equino-varus. The muscles in front of the tibia were wasted, the tibialis anticus and extensor longus responding to a faradic current which did not excite the peroneus longus. On the

left side there was a similar state of things but not to the
same extent.

Under massage, faradisation, manipulation,
and division of some tendons, this patient
recovered to a remarkable extent, being able in
the course of three years to walk with very
little aid.

Pains and extreme sensitiveness to touch are,
as I have said, of extremely frequent occurrence
in alcoholic paralysis.   It is interesting to note
that when recovery takes place and a second
attack occurs later on, the symptoms in this
respect may vary in the same individual.

A few years ago I saw a lady, aged thirty-seven, who
was unable to stand, and complained of dull aching pains
in her legs, which had been going on for three or four
weeks.   For four months she had been feeling weak, and
six weeks before I saw her she rather suddenly lost power
in her legs.   As she lay in bed she could draw up her
knees, but was quite unable to flex either foot.   The
anterior tibial muscles on the right side had entirely, and
those on the left side almost entirely, lost their faradaic
excitability.   There was no hyperalgesia to touch.

On inquiry it appeared that two years previously she
had slowly lost power in the legs " after a chill."   Next
day there were pains all over her, and on the following day she was unable to walk without help.   She was
then laid up with almost complete loss of power in the
lower extremities.   She suffered from excruciating pains
in the legs from the foot to the knee.   She also felt

numbness over the lower part of the trunk and down each thigh. Her hands at the same time were so weak that she could not hold things. She had slowly recovered after about twelve months. The character of the pains in this second attack she described as differing entirely from what she had suffered in the previous illness. They appeared then to have been of exceedingly acute character.

It may be worth mentioning in reference to this patient (whom I saw on one occasion only, she having been brought up to town for the purpose) that when the diagnosis was communicated to her husband he at first denied the possibility of its being correct, but on persisting in my opinion I obtained a history of an extremely large consumption of brandy, and an acknowledgment of the correctness of the view which had been taken.

One very important feature of this subject is that we may see a remarkable difference of severity in these cases, so that it is difficult at first sight to understand that they can possibly be examples of one and the same disorder. Dr. Broadbent, in his recent paper, brought forward several examples of a form of alcoholic spinal paralysis which he described as characterised essentially by insidious onset, progressive weakness, especially of the extensors on the forearms, giving rise to double wrist-drop, inability to stand, loss of knee-jerk, and retention of plantar reflex. The sensation was said to be unimpaired, except that in every case there was tenderness on firm pressure, and in one lancinating pains. There was œdema of the lower extremities. The symptoms gradually grew more intense, and death took place by asphyxia in consequence of paralysis of the diaphragm and intercostal muscles. In the case in which a necropsy took place no change could be detected in the spinal cord, the only organ which was allowed to be examined. I think the inference to be drawn from the clinical characters of these cases and the negative results of examination of the cord is that they were examples of multiple neuritis.

Dr. Wilks, Dr. Barlow, and myself, who took part in the discussion on Dr. Broadbent's paper, spoke of recovery as very common.* I remarked, too, that in more than one case that I had seen there was a history of a previous attack, from which under abstinence the patient had recovered, only to become again affected as the old habit was resumed. In an interesting work, ' Étude sur les Paralysies Alcooliques,' recently

---

* *Vide* 'Proceedings of the Royal Medical and Chirurgical Society of London,' February 12th, 1884, vol. i, No. 5.

published in France by Dr. Oettinger, the author speaks of the prognosis as "most grave," and does not conceal his surprise at the recorded experience of some English physicians who speak of patients recovering after abstinence from alcohol, to relapse when they again returned to it. This difference of experience would imply that a number of cases pass unrecognised, being referred to some other cause than alcohol; and this, it seems to me, may be quite as true probably in reference to the very severe as well as to the slight examples. An observer, whose experience of the disease had been derived from a certain number of fatal cases, might easily overlook the true origin of comparatively slight examples which came before him. On the other hand, one who had got to associate alcoholic paralysis with cases which uniformly recovered might not unnaturally sometimes fail to ascribe a rapidly fatal case of paralysis to its real source.

In writing some years ago upon the subject of lead palsy, I remarked that " the history of exposure to lead has often to be sought out by pertinacious inquiry after the diagnosis has been formed through other sources." And this, which is true of lead, is at least equally the case in regard to alcoholic paralysis, for here

the patient will often not only fail to give the
clue to the cause of his disorder but delibe-
rately, and sometimes with extreme dexterity,
put his medical attendant off the scent.

The art with which a secret drinker, especially if a female, will conceal her vice is well
known. Where there is an absolute conceal-
ment of all traces of alcohol and a dexterous
suggestion on the part of the patient of other
possible causes of the illness, it is not surprising
that the medical attendant is sometimes deceived.
This must evidently be especially liable to occur
in hospital practice on account of the difficulty
of obtaining accurate information as regards
the habits of the patient.

There is now enough of evidence from histo-
logical examination to show that in alcoholic
paralysis of the kind which I have described
the essential lesion consists in parenchymatous
neuritis of the peripheral nerves. It is evident
that, as a result of chronic alcoholism, more or
less extensive lesions may be expected to be
found in various parts of the body, especially
in the liver and intra-cranial membranes. But
there can be little doubt that the degenerative
changes in the peripheral nerves are the immediate cause of the paralytic symptoms.

In a series of examples published by Lancereaux in 1882, a careful examination of the fatal cases showed the spinal cord entirely free from pathological change. The roots of the spinal nerves also were apparently quite healthy, even under careful microscopical examination. The periphery of certain trunks, however, the posterior tibial and radial, presented the characteristic appearances of degenerative neuritis. Confirmatory evidence from other observers abounds as to the immunity of the cord and roots of nerves. With some reservation as to the nature of the neuritis, whether it is purely parenchymatous or of interstitial origin, they are evidently cases which come into the category of multiple neuritis.

I have already sketched before you the symptoms and course of multiple neuritis. These need not therefore be recapitulated, but it may be useful to refer to some of the clinical features of the alcoholic form of the disease as I have observed it. In the nature of things, more or less intellectual disturbance is present. The memory is especially weakened; the patients "talk freely," as Dr. Lettsom writes in the pamphlet before alluded to, " but of things that do not exist; they describe the presence of their friends as if they saw realities." A journalist told me of the articles he had just written and sent off to certain newspapers with an air of such *vraisemblance*, that it was difficult to credit the fact that he had not lifted a pen for three months. It may be found on inquiry that the patient suffers from nervous symptoms suggestive of incipient delirium tremens. I have remarked that pains and hyperalgesia have been, as a rule, extraordinarily pronounced. So, also, the degree of muscular atrophy seems to me to be frequently greater in this than in other forms of multiple neuritis. It is remarkable to see the extent to which in many cases the muscles of the legs and forearms are wasted. The muscular tissue seems to have almost entirely disappeared. This is especially to be seen in the extensor group, so that the feet, as the patient lies, drop helplessly forward. As has been already pointed out when dealing generally with the symptoms of multiple neuritis, the varying degree of muscular atrophy in a limb may easily give rise to contracture of rigid character. You may thus at one stage of the disease find a patient lying in bed with powerless, wasted, and flaccid limbs—the feet and hands, as mentioned, helplessly dropped; and in another stage find the same patient with the tendo-Achillis rigid, the foot unable to be brought into dorsal flexion by strong passive movements, the hamstring muscles contracted, whilst the hands show the claw-like character described by Duchenne.

The pains and inordinate sensibility of the skin may, I believe, be altogether absent in alcoholic paralysis, as happens likewise in some cases of multiple neuritis of non-alcoholic origin. The following case, which recovered in the hospital, was a striking example of this :

A single woman, aged thirty-one, was admitted into hospital with loss of power in all four limbs, but especially in the legs and feet. She could not stand even with help, the legs doubling up under her when the attempt was made. Lying in bed she could straighten her legs and flex the knees. She could also move the feet a very little, but the toes not at all.

As regards the upper extremities the loss of power in the hands was very great in proportion to the strength of the arms—indeed, the arms did not seem at all affected, as tested by exertion against resistance. The grasp of the right hand was 10° of the left 0. The sensibility of the legs below the knees and the feet was impaired, and also that of the hands. The plantar reflex was obtained with difficulty; there was no knee jerk on either side.

The patient was a very fat woman with sallow complexion and unhealthy aspect. Her feet and the lower part of the legs were œdematous, pitting on pressure, and so were the hands and forearms, though to a less extent.

The heart was free from any signs of disease; there was no albumen in the urine, nor was there any ascites.

Owing to her obesity the size of the liver could not 'be made out.   Her account of her illness was that six months previously she had noticed gradual loss of power in her legs, and after six weeks was obliged to leave off her work, which was of a mechanical nature.   The history she gave was indefinite, but apparently she had suffered from " pins and needles " in the feet and legs, and no complaint was made of pains.   She had not had any pain in the back, The swelling of the legs had occurred for some time, but had increased during the month previous to her coming under observation.   The hands and forearms had only shown swelling for about a week ; the want of power in them had been of about one month's duration.   There was no difficulty in regard to the bowels, but she had latterly been obliged to be quick in obeying the call to urinate.

This patient began to improve shortly after her admission, and in a month could take about a dozen steps with slight assistance.   The grasp of the left hand then marked 8°, that of the right only 10°, as on admission.   She could move the toes a little and the sensibility of the lower limbs had returned except over the œdematous feet.

After another month she could walk the length of the room alone, although in an ataxic fashion.   The knee-jerk was still absent.

After a few months she entirely recovered and returned to her employment.

Whilst in hospital she lay in bed for the first month, and later on some electrical treatment was employed. Iodide of potassium was given.

She had never had rheumatic fever, but a year previously appeared to have suffered from pains about some of her joints.   It is right to say that no reliable information could be obtained as to her habits.   From her general appearance, however, the fact that she had long been affected with morning retching and sickness, with aversion from food, and that immediately previous to the commencement of the paralytic symptoms she had suffered from jaundice, it was judged probable that there had been an excess in alcohol. Whilst in hospital she had no opportunity of continuing the indulgence, and, as I have said, her improvement was rapid and recovery complete.

This patient showed in a marked degree the characteristic excess of paralytic symptoms in the feet and hands as well as the vaso-motor paralysis causing œdema.

The case was one, there can be little doubt, of multiple neuritis, which was of alcoholic origin.

The absence of knee-phenomenon is so common in these cases that we may almost confidently expect to find this symptom. It will now and then happen, however, that we may find the knee-reflex not only present but somewhat exaggerated. I do not see how to explain this as a result of neuritis, and as, I believe, it is only in alcoholic examples that the anomaly is observed, it may be due to interference with the inhibitory influence of the cortex cerebri caused by the action of alcohol.

In the early stage of a case related by Dr. Oettinger, a highly characteristic one of this class, the knee-phenomenon is described as " notably exaggerated," and it is remarked that tickling the soles of the feet gives rise to " very extensive reflexes." In the sequel of that case the knee-phenomenon disappeared.

A case which has lately been under observation in hospital is an example in point. It presents too, some interesting features, which induce me to bring it before you in some detail. The patient, previous to admission, had been under the care of my colleague, Dr. Ormerod, who had diagnosed the case as one of alcoholic paralysis.

A working man, aged thirty-eight, experienced a feeling of stiffness in both legs, especially in the calves, with aching in the spine and shooting pains in his limbs. When these symptoms had lasted about two months he felt both his hands become weak and numb, and, according to his own description, "as though they did not belong to him." If he rubbed them together they would tingle. He was admitted into hospital seven months from the commencement of his illness. At that time he could not walk except with great difficulty and with a very peculiar gait. The legs were widely separated, the body thrown forward, and the ground was only touched with the anterior part of the foot, the heels never coming down. He could not stand still in this "equine" position, but was obliged to hurry forward. When his eyes were closed he was very insecure.

The tendo-Achillis on each side, and especially on the left, was tense. When lying down he could perform all movements with his legs, but the power of dorsal flexing and inverting the right foot was most imperfect, whilst on the left side these movements could scarcely be performed at all.

Examination with electric currents in February showed loss of faradic excitability in the muscles of both legs below the knee, and in the vasti of the left thigh. There was slight reaction to strong induced currents in the vasti of the right thigh.

The knee-phenomenon was well marked on both sides, especially on the left. Plantar reflex was well marked and produced more vivid reflex movements in the opposite leg than in the one touched. This was especially to be remarked of the left leg when the sole of the right foot was tickled. The abdominal and epigastric reflexes were well marked, especially on the left side. Tactile and painful sensibility were not apparently affected, but he complained of numbness in the feet and hands. The sciatic nerves and the anterior tibial nerves in their lower third were tender on pressure. Many points of tenderness were to be found over the spinal column, especially about the lower dorsal region.

The arms looked thin and wasted but presented no marked loss of reaction to electric currents. The grasp of each hand was very weak, the dynamometer giving 40° for the right, 45° for the left.

Patient had a half-jaundiced appearance; his liver was large; there was no albumen in his urine. For many months he had suffered from diarrhœa, morning sickness, and loss of appetite. He had been in the habit of drinking a great deal of beer since he was twelve years of age. On an average he had taken ten pints of beer daily, and sometimes whisky as well. He had twice suffered from delirium tremens.

This patient was treated by confinement to his bed for some months, total abstinence from alcoholic liquors, and the constant current was applied to his legs. In May a very distinct improvement had taken place. He could walk without hurrying, and the tendo-Achillis was much less tense, the sole of the foot being now able to be placed on the ground. At the end of May the tibialis anticus muscle on each side reacted fairly to induced currents, the extensor longus digitorum acted slightly to very strong currents, the peroneus longus and the vasti showed normal reaction to induced currents. In June he could dorsal-flex and invert either foot. A few weeks later he was discharged quite recovered, his gait having ceased to present any peculiarity.

The absence of the knee-phenomenon which is so generally observed in all forms of multiple neuritis, coupled with the lightning pains so often experienced by the patient, may be strongly suggestive of tabes dorsalis. This resemblance is sometimes increased by the occurrence of a notable amount of ataxy. In the case of my patient, T. O—, there were sharp pains, sudden, and of momentary duration, like a knife stab in the right thigh and knee, his gait was ataxic, and he said the ground did not feel natural to him. His legs seemed to spring under him. The knee-phenomenon was absent. At a certain stage of his illness the superficial resemblance to a case of tabes was very striking. A notable point of distinction was to be found in the behaviour of the muscles to electrical currents. It is well known that in tabes there is essentially no change from the normal condition in this respect. In certain cases no doubt the anterior grey matter of the cord may become invaded and cause muscular wasting with loss of faradic excitability in limited parts, but this is quite, as it were, an accidental complication, and is not an essential part of the disease. Now, in my case of multiple neuritis there was very slight reaction to strong induced currents in all the muscles of the lower extremities, and almost total absence in the interrossei and thenar muscles of the right hand. This of itself, to say nothing of other differences, was sufficient to distinguish the case from one of tabes.

Dr. Hadden has drawn attention to the occasional occurrence of nystagmus in these cases. The observation is an important one, as the ataxy gives sometimes a *primâ facie* resemblance to disseminated sclerosis, which the presence of nystagmus would be liable to support.

A case was lately published by Dr. Churton, of Leeds, which has an interesting bearing on this subject. It was one in which ataxy and other nervous disorders occurred in a syphilitic patient.

The patient was a groom, twenty-four years of age, accustomed to drink rather heavily of beer and spirits. When admitted into hospital he was unable to walk, and could stand with great difficulty even when supported on each side, anxiously looking down all the time at his legs, and so tending to fall forwards, being unable to balance his trunk upon his lower limbs. He had also lost power of co-ordinate movements in his upper limbs, could not cut up his food nor even pick up a match without great trouble, nor tell by the sense of touch what the match was. His grasp was not very feeble; he complained of numbness and pins and needles in his fingers. His replies to questions were hazy, and he seemed dazed. There was no tenderness nor swelling of bones or joints and no special wasting of muscles. The plantar and patellar reflexes were absent; the cremasteric and abdominal present; the pupils, optic discs and retinæ were normal. He had had a chancre sixteen weeks before admission, followed by sore-throat and rash. Five or six weeks before admission the toes of his right foot began to be numb and tingling, he staggered in walking, and the numbness spread up the leg, Then the left foot and leg became affected and he could not stand with the eyes shut or in the dark. At this time he was still drinking rather freely. About a month later he was quite unable to walk, and a few days after that his hands became numb and weak.

He was submitted to a vigorous mercurial treatment, and recovered so rapidly that he was discharged from hospital on January 21st, having been there a little more than a month. A month later, it is noted, the reflexes were normal.

Dr. Churton remarks that although the prominent features of this case were those of locomotor ataxy or posterior sclerosis it cannot be taken as a simple case of that disease. I should myself feel disposed to think that this case was one of multiple neuritis. Where the history points both to syphilis and alcoholism, it is difficult to allot the part played by each in the production of the disease in this instance, but it is probable that alcohol was largely concerned. The absence and return of patellar tendon reflex is of constant occurrence in cases of multiple neuritis. I have never seen it restored (after being lost) in a case of well-marked posterior sclerosis. The mental obfuscation which is described has no association with that disease, whilst it might be expected in a case of multiple neuritis from alcohol. The absence of plantar reflex also, so far as it goes, favours the view that this was the nature of the case.

Déjerine, in France, has drawn attention to cases in which pains, inco-ordination, absence of knee-phenomenon, and anæsthesia have produced a striking resemblance to tabes, and in which after death no lesion of the cord was found, but there were degenerative changes in the peripheral nerves. He has suggested for these the title of " neuro-tabes périphérique." Facts of this kind have to be borne in mind ere

we conclude of a case marked by the characteristic symptoms described that it is one of sclerosis of the posterior columns.   In Déjerine's
cases I cannot help thinking that alcohol was
the important etiological factor.

Considering that the toxic influence of alcohol
must be brought about through the medium of
the circulation, it is not surprising that the
upper as well as the lower extremities should
be affected in cases of alcoholic paralysis.
Indeed, it might be anticipated that the effects
would display themselves equally upon all the
voluntary muscles of the body.   But this is not
the case.   It is upon the lower extremities that
the brunt of the mischief falls.   They usually
suffer the most, and may possibly, perhaps, be
occasionally alone affected.   But I am disposed
to think that their immunity is not nearly so
great as has been supposed, and that careful
observation would show that in cases where
the patient only complains of loss of power in
his legs, the arms are also, though to a less
extent, likewise affected.   The patient's attention is apt to be so engrossed by the preponderating disorder in his lower extremities that
he takes little or no notice of the weakness in
his hands.   An observation which I made many
years ago in a case of lead-poisoning very much

struck me.   Although the patient only complained of one arm and one leg (which were
manifestly paralysed), and asserted that there
was nothing wrong with the other extremities,
I found in the muscles of the latter a very
well-marked decrease of faradic excitability.
I have also many times noticed a similar condition in cases of infantile paralysis.

In general terms, it may be said that just as
in a case of lead paralysis we expect to find
wrist-drop so in a case of alcoholic paralysis we
look for dropped feet.   I would go further
even, and say that if we meet with a case of
dropped feet—a paraplegic condition affecting
with marked preponderance the anterior tibial
group of muscles—we should be on the alert to
inquire respecting the possibility of alcohol
being a cause.   Let me not be misunderstood.
The existence of foot-drop is not alone a proof
of habits of excess, but the symptom is so extremely frequent in cases of alcoholic paralysis
that we should be wanting in our duty if we
failed to bear this in mind, and direct investiga-
tion accordingly.   This is of course a delicate
matter, and on more than one occasion I have
observed a look of somewhat indignant surprise
on the face of the medical attendant of whom
the inquiry has been made.   But we have no

more right to omit the inquiry than we should have to avoid examining into the possibility of lead poisoning when a case of dropped wrist comes under our observation. It is especially when we find not only the extensors of the feet but those of the hands paralysed, and also when there are some sensory disturbances as well as motor, that we shall do well to bear in mind the possibility of alcohol being at least a factor. Where careful observation shows that the lower extremities are alone involved, the upper extremities being quite normal as regards strength, sensibility, and electrical reaction, it will usually, I think, be found that the influence of alcohol may be put out of the question. It is evident that there is little likelihood of the effects of alcohol being limited to certain extremities; but, as I have said, it is very common for the legs to show the disorder before the arms—and supposing that abstinence begins to be practised at this point of time it is perhaps conceivable that the latter might escape. But this is probably very uncommon.

I am not able to explain the greater tendency of the lower extremities to suffer in this affection. It is an interesting circumstance that a similar proclivity for the lower extremities to

be most affected (sometimes indeed exclusively
so) is shown, as I have before remarked, in the
case of the endemic disorder, Beriberi. But it is
not only in connection with alcohol and Beriberi
that this preponderance is observed. Several
cases have fallen under my observation marked
by characteristic symptoms of peripheral neuritis
which have been entirely confined to the lower
extremities.

The following is a striking example :

A gentleman, aged forty-nine, consulted me on account
of numbness of the feet, a difficulty in standing with his
eyes shut, and a feeling as though he were walking on
bladders. His gait was not ataxic. He had been invalided
from abroad on account, as he told me, of incipient locomotor ataxy. His pupils were small ; the knee-jerks were
not elicited by blows with the percussion hammer. The
primâ facie resemblance to tabes was, I need scarcely say,
very strong. Further examination showed the following : 1st. The patellar tendon-reflex was brought out in
perfectly normal fashion when the side of the hand was
used for striking the ligamentum patellæ. 2ndly. The
pupils, though small, contracted to light. 3rdly. The
power of dorsal-flexing the feet was much impaired. 4thly.
There was marked anæsthesia in the lower half of each
leg. The wire-brush, with the strongest induced currents
applied over the parts shaded in the diagram, Fig. 4, was
scarcely felt at all in the feet, but gradually evoked sensation
as the leg was ascended, and above the line of shading the
pain of application became intolerable. About the upper
part of the anæsthetic districts an ordinary touch with the

finger-point was described as feeling like "the pricking of an electric battery." The patient complained of a con-

FIG. 4.

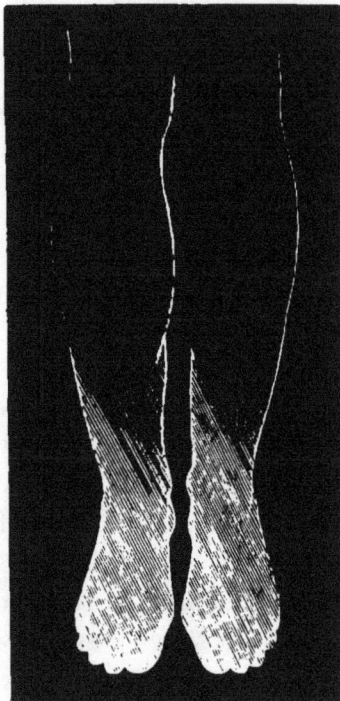

stant aching and numbed sensation in the soles of his feet, which was not increased by walking. The sensation was like that of the feet being very cold indeed in the boots on a cold day. Electrical testing showed on each side considerable loss of faradic excitability in the extensor longus digitorum and tibialis anticus, the peroneus longus being apparently normal. The response to interrupted galvanism was not increased. 5thly. The patient had never had any sudden shooting pains, but there had been for some time a dull heavy pain in the soles of the feet, along the shin-bone, and the back of the leg. This was constant. It was difficult to obtain a very distinct account of the mode in which the symptoms had commenced. So far as I could judge there had been extremely painful swelling of the right foot six years before, which had confined him to his bed for two months, and terminated in an abscess at the root of the little toe. Two years later he appears to have had a collection of matter in the left foot which discharged near the little toe. Since that time he had a sensation of numbness in the sole of the left foot, and for the past nine months the numbness had become acute in both feet. There was no history of syphilis nor of gout, and I do not think there had been alcoholic excess sufficient to explain the symptoms. Under the employment of faradic currents with a wire brush a great improvement took place in the sensibility of the skin of the feet. Unfortunately at this point I lost sight of the patient.

In this case there would seem to have been chronic neuritis affecting the terminal branches of the popliteal nerves, with the exception of the supply to the peroneus longus and brevis. The cause of the neuritis, however, is by no means clear.

Another patient was a man of middle age who considered himself to be gouty. When I saw him he had paralysis of the anterior tibial group of muscles in both legs, the feet being dropped in consequence. He was covered from head to foot with copper-coloured papulo-squamous eruption, which had appeared about five months previously. It was whilst under mercurial treatment for the eruption of the skin that he suddenly one night lost the use of the left foot, and this was followed by a corresponding failure of power, but not to the same extent, in the right foot.

On electrical examination I found in the left leg no reaction to either current in the anterior tibial muscles, and so much cutaneous anæsthesia in the lower half of this leg that the strongest faradism was not felt. In the right leg there was also some anæsthesia in the same situation. The extensor of the toes required very strong induced currents to cause contraction; the tibialis anticus was less abnormal.

A fortnight later there was slight negative pole closure contraction with the strongest galvanic current in the tibialis anticus of the left side.

A couple of months later I failed to get any contraction to either current in the muscles of the left leg in front of the tibia on the left side. The patient described a numbed, dead feeling in the feet. He could not tell what he was walking upon. Both ankles swelled at night. I found on two or three occasions that the left extensor carpi radialis muscle was a good deal less excitable to faradism than the right. He was treated with iodide of potassium. I lost sight of this patient, but a few months later learned that he had quite recovered.

The following case appears to me one closely resembling those described as dependent upon alcoholism, but this cause can be absolutely excluded. I am unable to suggest a cause for the chronic neuritis which evidently exists.

A female, aged sixty-three, of active and temperate habits, had enjoyed good health till 1880, when she complained of general weakness. In 1881 numbness was noticed in the feet and legs, and she began to be readily fatigued in walking. These symptoms increased slowly but steadily during the next twenty-four months until walking for more than a few steps became impossible. There was distinct loss of power in the muscles lying on the left tibia, the foot being dropped. Little or no pain was felt at any time except when the affected parts were handled or pressed. The following parts were acutely sensitive, even to gentle pressure: dorsal and inner surface of the feet, the front and sides of the legs and of the thighs, the lumbar and sacral regions of the trunk; to a less extent the soles of the feet, the middle of the back, buttocks, and flanks. This sensitiveness was most marked on the left side, especially in the lower part of the left leg and foot. There was considerable œdema of the legs from the knees downwards, none elsewhere. There was no albuminuria or cardiac mischief.

With the exception of occasional attacks of something described as resembling petit mal, the patient's general health has remained good, notwithstanding that she has been obliged to give up all exercise, and is confined to her couch.

Such was her history given me when I saw her about a year and a half ago. I found the power of flexing thighs on trunk and legs on thighs unimpaired. She could dorsal-flex the right foot fairly well, but the left very imperfectly, and there was marked "dropping" of the left foot. Heat and cold produced normal impressions. The sensitiveness to touch and handling I have described above. The plantar reflex was absent on the left side. The knee-jerks were fairly marked in each leg. There were no lancinating pains, and no marked atrophy except perhaps in front of the left leg, but this is obscured by œdema. The toe-nails were darkened in colour and some were roughened, especially those of the great and second toe of the left foot.   I thought the response to induced currents lowered in the left leg and fairly good in the right one, but the œdematous condition interfered with exact observation.

At the present time, I am informed by her medical attendant, the legs are as morbidly sensitive as ever, the right being equally so with the left, but the œdema is less, especially in the left leg.   There is no loss of muscular power in any fresh place, and a distinct gain in the left anterior tibial region, as manifested by greatly improved power of dorsal-flexing the ankle-joint, and a little more power of extending the toes.   This improvement has coincided with the application of the constant current.

LECTURE III.

Peripheral neuritis in enteric fever—After dengue—Two cases of peripheral neuritis after malarious fever—Alcoholic history in each—Doubtful etiology—Phases of alcoholic paralysis—Diphtheritic paralysis—Resemblance of symptoms to those of multiple neuritis —Tenderness of nerves and muscles—Absence of knee phenomenon—Varying condition of electric excitability in diphtheritic paralysis — The ataxy after diphtheria liable to be mistaken for tabes—Morbid anatomy—Diphtheritic paralysis probably very often dependent upon peripheral neuritis — Comparison between symptoms of alcoholic and diphtheritic paralysis—Diagnosis of multiple neuritis.—Prognosis—Treatment.

It is necessary to bear in mind that peripheral neuritis occasionally occurs in the sequel of some febrile affections. Besides diphtheria, of which I shall have to say more presently, other specific febrile disorders are apt to be followed by paralysis, which is sometimes more or less general and at other times strictly localised. Dr. Ross has collected a number of published examples in which there has been localised paralysis of peripheral origin in the sequel chiefly of enteric and relapsing fevers. I was consulted a few months ago by a gentleman who was suffering from paralysis of the serratus magnus of the right side, which had occurred whilst he was convalescing from an attack of enteric fever, acquired during a tour in Italy.

It often happens that owing to the prostration and unconsciousness of the patient, the characteristic symptoms of neuritis are obscured and a localised paralysis is alone noted, the immediate cause of which there is no evidence to explain. But in other cases, violent pains and other disturbances of sensibility accompanying paralysis, with muscular atrophy, and partial or complete removal of faradic excitability, point conclusively to a peripheral neuritis as the pathological condition.*

I have lately seen an officer of the army, formerly on service in the tropics, who has had for the last year and a half paralysis of the flexors of his left foot. There is atrophy and loss of electrical excitability in the tibialis anticus and extensor longus digitorum muscles

---

* Since this lecture was delivered an important contribution to our knowledge of this subject has been published by Pitres and Vaillard in the 'Revue de Médecine,' December, 1885.

coupled with some anæsthesia of the corresponding surface.   He has never suffered from syphilis, and the only antecedent of importance is a severe attack of dengue.   This occurred, it is true, more than two years before he began to get pain in the left knee, by which the paralysis was ushered in, but meantime he had suffered now and then from an attack of rigor like the recurrence of ague.

The possibility of paralysis from peripheral neuritis occurring in the sequel of specific febrile affections may give rise to a curious difficulty of diagnosis.

A few years ago a gentleman brought me a letter from his medical attendant abroad, from which I will quote :

"The patient, after a severe attack of pernicious malarious fever, became paralysed in both legs.   When he came under my notice he was unable to move his legs or raise himself in bed.   The sensibility was quite normal. The electric contractility of the muscles of the legs was much diminished, the liver and spleen were normal, the urine contained an unusual amount of phosphates.   The memory was very weak.   Under the treatment the patient improved rapidly so that he can now walk without any help, can ride, and shows a great facility of remembering facts and dates.   The motion of the ankle remains still injured.   With his eyes closed patient can neither stand nor walk.   The treatment consisted in the application of weak currents along the spine and local faradisation of the muscles of the leg twice a day, tonics, nitrate of silver, strychnine.

"I advised Mr. — to visit a hydropathic establishment, to undergo there a cure and continue, if necessary, the faradisation. As he is very fond of liquors, and as these had always a very injurious influence upon his memory and the steadiness of his walking, I only allowed him two pint bottles of beer and a few small wine-glasses of Madeira a day. At night he received a small quantity of whisky with water. The disease has surely been caused by the fever, serous exudation being the secondary cause of the paralysis."

When I saw this patient he had quite recovered, and the diagnosis is necessarily therefore uncertain. How far the attack of pernicious malarious fever was concerned in the production of the nerve changes, it is impossible to say, but the symptoms are quite consistent with those of the multiple neuritis which is connected with chronic alcoholism.

A few years ago (the date is purposely omitted) I saw a female patient, aged thirty-three, who came to me with the following written account from her medical attendant in the tropics. After stating that during her residence in the country she had had several attacks of tropical dysentery and had been the subject of some obscure biliary affection at the close of which he saw her, he went on to say, "The symptoms called 'biliousness' had then disappeared and had given place to debility, anæsthesia, and an incipient paralytic condition of the upper and lower limbs, in many respects resembling locomotor ataxy. There was also and had been for two months cessation of the menstrual flux. In addition there was marked loss of memory, and more than once I observed her conversation incoherent. Under phosphorus, strychnine, iron, galvanism, friction, this condition of incipient paralysis slowly departed, and, with the exception of reduced strength and impaired memory, she was nearly well till the end of February, 18—. At that time the menses had again ceased, irritability of stomach, flying pains and numbness of limbs, with utter inability to walk had again returned. Though she could move her limbs when lying in bed, and though sensation was but little at fault, she could neither stand on her legs nor walk a step. She complained bitterly of pain and tingling and numbness in both arms and hands, and the co-ordinate power of the small muscles of thumb and fingers was greatly impaired, the left more than the right." She was treated with counter-irritation over the spine, iodide of iron, later by galvanism and phosphide of zinc with strychnia. "She was allowed a very nutritious diet of milk, beef tea, bread, eggs, but no alcoholic stimulant. After three or four weeks she had improved, and a little later the menses reappeared." The opinion expressed by the medical attendant was that her disease was due to anxiety from severe bereavement, coupled with residence in a malarious tropical country.

This patient arrived in England, and came to me two and a half months from the beginning of the relapse described. Her complexion was sallow and conjunctivæ yellow. She had recovered power to a great extent. On starting for England she had been unable to dress herself. Now she could do anything with her hands, but delicate movements, as in buttoning things, were not well performed. She could walk a moderate distance. The knee reflex was absent on each side. Electrical examination showed a great reduction in excitability of the muscles in front of each leg. This lowering of excitability obtained both in regard to the galvanic as to the induced current.

This lady came from a country where malarious fever was rife, but I could get no evidence of any characteristic intermittent symptoms. Nor did I learn anything with regard to the patient's habits. The case was evidently one of multiple neuritis, and from the symptoms described, so characteristic of peripheral changes, besides the occasional incoherence and loss of memory, pointing to intracranial lesion, it seems probable that alcoholic excess had been at least an important factor in the case.

One occasionally meets with alcoholic cases in which there is paralysis to a considerable extent which has escaped the notice of the patient's friends. The patient (usually a female) lies in bed in such a muddled condition of mind that she does nothing for herself, and takes sustenance at the hands of others. When moved from side to side in bed, or when her limbs are touched, she cries out with pain. It is not at all uncommon to find this condition referred to rheumatism or gout. In some cases, as an explanation of the mental condition, one is told that the rheumatism or the gout has "flown to the brain." Although evidently there is a great probability of persons who indulge to excess in alcohol being affected with gout, yet I am disposed to think that much more often this condition is due to peripheral neuritis from alcohol.   The persons who suffer in this way do not drink wine or beer—the liquors which tend so much to induce gout— but brandy or gin.   In such cases as I have described there will often be an amount of muscular wasting of the extremities, not to be explained by mere emaciation.   If examination be made, it will very likely be found that the knee-jerks are absent, and the faradic excitability in the muscles greatly reduced or lost.

In a case which I saw in consultation some time ago there was intense pain of a neuralgic character in both lower extremities, with dropping of the feet and absence of knee-jerk.   A curious question arose in reference to diagnosis.   The patient had been operated upon for a carcinomatous tumour (unconnected with the spine) a year or more previously.   The symptoms pointed to secondary growth in the lower part of the spinal column, causing neuritis in some portions of the cauda equina.   But there was also a distinct history of long-continued and great alcoholic excess.   Examination showed that whilst faradic excitability of the muscles of the lower extremities was greatly reduced, that of the muscles of the arms was quite normal.   This being the case, I came to the conclusion that the paraplegic symptoms were due to malignant disease. It was unlikely that such severe symptoms of neuritis would have been exclusively confined to the legs had

alcohol been the cause.   At least one would have expected
to find some change in the electrical reactions of the upper
extremities.   The opinion given was in accordance with
this observation, and was, I believe, justified by the
sequel.

Amongst the toxic influences apparently
liable to give rise to multiple neuritis, besides
alcohol, syphilis, and the essential cause of
Beriberi, whatever that may be, I have mentioned lead and diphtheria.   There are circumstances connected with the question of the
pathology of lead-poisoning which appear to
place it on a somewhat different footing from
the other varieties described ; and this, together
with the narrow limits of my space, renders
its discussion on the present occasion imprac-
ticable.   I will therefore conclude this part of
my subject by a few remarks upon diphtheritic
paralysis.   The symptoms of diphtheritic paralysis are so well known that I need not trouble
you with any systematic account of them.   My
intention is to refer chiefly to those clinical
features which show that the disease ought
probably to find a place alongside of the other
forms of paralysis essentially dependent upon
peripheral neuritis which we have been considering.

Last year I unfortunately had the opportunity

of observing very constantly a case of diphtheria which was followed by paralysis.

The patient was a boy nine years of age, who was severely attacked with diphtheria on December 16th. The temperature, which was nearly 105° F. during the first twenty-four hours, then declined to a little over 100°, running up to 102° on the sixth day coincidently with the appearance of albuminuria. On the thirteenth day the throat had recovered, the albuminuria continuing till the twentieth day.

On the thirteenth day the patient spontaneously complained of much tenderness on the inside and back of each thigh and on the fifteenth day pressure over the sciatic, posterior, tibial, and median nerves was decidedly painful. Throughout his limbs handling and gently squeezing the muscles appeared to cause considerable discomfort. The knee-jerks at this time were quite normal.

About the twenty-third day he began to complain of difficulty in reading, and in a few days he could not read at all. Mr. Macnamara, who kindly examined him, found almost absolute inability to accommodate. On the ninety-ninth day Mr. Carter found that the left eye had recovered its accommodation, that of the right still remaining paralysed. This also recovered later.

On the thirty-eighth day he complained of a peculiar feeling at "the top of the throat," and some water which he drank came for the first time through his nostrils. He spoke with a nasal twang. He fidgeted much and felt as if his legs would give way, and said he could not stand still. At this time he never passed a day without complaining of pain referred to the epigastric region. The pupils were dilated. They responded to light. There was no optic neuritis.

On the fortieth day, in going downstairs, he staggered a good deal, and had not complete control over his feet. As he himself expressed it, "he wanted to take one step, but took two." The knee-jerks, which had been examined daily from the time of subsidence of the throat symptoms, were still quite normal. It was not until the ataxy of gait had lasted nearly three weeks that the knee-jerk became rather suddenly very feeble. Next day (the sixty-first of his illness) the phenomenon was absent on the right side and very small on the left, and on the following day, the sixty-second of his illness, there was not a trace of tendon reflex in either leg. Anticipating a little for the sake of clearness, I may here say that the knee-jerks returned about the first week in June—*i. e.* about the 170th day of his illness, They were absent, therefore, about 108 days. I had examined him daily until April 21st, when the lad went into the country. He had learned perfectly to test his knees for himself, and reported with much satisfaction the date of return of the reflexes.

It was especially from about the end of January till the end of March that the boy showed great awkwardness of gait and clumsiness in the use of his hands. He could only walk at a slow pace, and was very soon fatigued. He would now and then, with boyish instinct, try to run, but in this he quite failed. He could not perform the movement. At times, in walking, he would suddenly come to the ground, having tripped or in some way missed his foothold.

In the beginning of April, whilst the knee-jerks were still completely absent, he had so much improved in the power of his limbs that he could walk, run, and join his brothers in play, and this progressive improvement continued uninterruptedly until his recovery. The nasal character of his voice and the regurgitation of fluids, which had commenced on the thirty-eighth day of his illness, were at no time very strongly marked symptoms. It was only occasionally that he had difficulty in swallowing fluids, and these symptoms ceased to trouble him at all after lasting about a month. At the time when the voice first became nasal in character it was noted that in speaking the lips were moved much more than was natural. There was, indeed, an ataxic over-action of the muscles about the mouth. A note to this effect was made on January 24th. It is interesting that within a week a complete change in this respect had taken place. On January 31st the muscles of the lips are described as "very indolent." On February 4th, again, along with mention of regurgitation of his coffee and the nasal tone of his voice, it is remarked, "Very little movement about the muscles of the mouth."

On the sensory side a prominent complaint was of pain in the stomach, of which he had early spoken, and which lasted longer than any other symptom. It occurred especially but not exclusively after eating, and was not accompanied by vomiting. I frequently pressed upon the nerve-trunks of his limbs in different situations, and satisfied myself that there was very distinct tenderness in the trigeminal, musculo-spiral, median, ulnar, sciatic, and popliteal nerves, as well as in the epigastric region.

No change in the electrical reaction of his muscles was at any time apparent. Although in this instance, and several others that I have observed, there has been no alteration in the electrical reaction, this is by no means constantly the case, and I have notes of several in which the excitability was distinctly lessened in some muscle-groups of the extremities.

The following are examples :

A gentleman was affected with diphtheria in July, 1872. When he consulted me in October of the same year he was suffering from general paresis. He walked with difficulty, the feet being lifted laboriously, and his grasp was very feeble. There was great weakness indeed of all the voluntary muscles. The pupils were dilated, and accommodation lost.

It seemed that three weeks after his convalescence from diphtheria he began to find difficulty in swallowing, fluids returning through his nostrils. Then he observed dulness of sensibility in the arms and legs, and his articulation became nasal, "like a Yankee," as he described it.

I found on examination that the degree of excitability by the induced current differed very much in various sets of muscles of his extremities.

A girl, aged eight, was seen by me on January 6th, 1876. She had suffered from diphtheria in the preceding July— an extremely bad attack, lasting five or six weeks, in which she was not expected to survive.

A month or six weeks after convalescence she had measles. Just before this fluids had begun to run from her nostrils. She walked clumsily, and her fingers became weak, so that she could not grasp. Her eyes looked peculiar, and one was fixed so that she saw double. The voice was nasal.[1] This was after the measles, about October. I found the faradic excitability distinctly lessened in some muscle-groups of the extremities.

The ataxy of gait to which allusion has been

made, and which is so frequently seen in the sequel of diphtheria, is precisely of the kind which occurs in multiple neuritis arising from other influences, such as alcoholic, syphilitic, and also in certain unclassed forms. It never, so far as I have seen, acquires the pronounced character of the ataxy which is so often associated with sclerosis of the posterior columns of the cord. We do not see in these cases the wild flourishing about of the legs, with stamping action, so characteristic of the latter disease. The ataxy in multiple neuritis is probably due, I think, not to any affection of the posterior columns of the cord, but to the want of harmony in the strength of various muscular contractions owing to the varying amount of the lesion of peripheral nerves.

I long ago drew attention to the chance of diphtheritic paralysis being mistaken for tabes dorsalis, of which several instances have come before me. When the electrical reaction of the muscles is unimpaired this is peculiarly liable to occur. The condition of the pupils should be a help towards differentiation. In tabes, as was first pointed out by Argyll Robertson, the pupils are very apt to lose their capacity for contracting to light whilst retaining the power of contraction during an effort at accommodation. In diphtheritic paralysis, so far as I have seen, the pupils retain the power of contracting to light. How they behave in an effort at accommodation I cannot say, and several ophthalmological friends to whom I have applied have not been able to tell me, but have kindly promised to observe the point. In view of the peculiar affection of accommodation in this disease, it would be interesting to know whether there is normal contraction of the pupil during convergence for accommodation. The tenderness of the trunks of nerves which was here conspicuous was long ago observed by Dr. Greenhow. In a paper read before the Clinical Society of London upon " Diphtheritic Paralysis," in 1871, he suggested the existence of neuritis, having been struck by the tenderness of the sciatic nerve in a case of this kind. He referred to the fact that Mr. Hulke in one example had discovered optic neuritis. In the cases of multiple neuritis which I have seen, including those of the alcoholic form, the optic nerve has been unaffected; but in one patient who, I believe, suffered from the non-alcoholic form of the disease (though the diagnosis is not absolutely sure) there was optic neuritis. It occurred also in a case related by Strümpell.

Opportunities of examination after death in

cases of diphtheritic paralysis are comparatively rare. The results in such cases as have been recorded are at first sight conflicting. Changes of an atrophic character in the ganglion cells of the anterior horns of the spinal cord have been described by Vulpian, Déjerine, Abercrombie, and others. The year before last Dr. Percy Kidd brought before the Royal Medical and Chirurgical Society facts in support of the view that diphtheritic paralysis is founded on a distinct anatomical lesion of the spinal cord. Specimens from a fatal case showed alterations in the shape of the large motor cells in the anterior horns and changes in their cell protoplasm. The affected cells were as a rule more or less globular in shape and devoid of processes. In certain regions of the cord corresponding to the distribution of the muscular paralysis during life there was apparently a numerical atrophy of motor cells. The changes were purely parenchymatous. There was no distinct affection of the neuroglia. I had the opportunity of examining Dr. Kidd's specimens, and so far as I am capable of judging, convinced myself of the accuracy of his observations. In the case in question no examination of the nerve-trunks had been possible.

On the other hand, many investigations

support the opinion that the cause of diphtheritic paralysis lies in a lesion of the peripheral nerves. Charcot, Vulpian, Lorain, Lépine, Lionville, Leyden, and Meier have contributed testimony in this direction. Changes also in the vascular system and its contents have been noted by Buhl, Oertel, Mendel, and others in the form of hyperæmia, capillary hæmorrhage, and thrombosis.

A recent and important contribution to our knowledge of the subject has been made by Mendel, who had the opportunity of examining a case which terminated fatally only ten days from the commencement of paralytic symptoms appearing in the muscles of the eyes and of the extremities. In the right orbit there had been paralysis of the superior and inferior rectus muscles, and paresis of the external rectus; in the left, paresis of all the muscles. Microscopical examination showed the ganglion cells of the deep origin of the oculo-motor nerves unusually large, as if swollen. The most striking changes, however, were found in both oculo-motor nerve-trunks, in the right more than in the left, and consisted in signs of interstitial and parenchymatous neuritis in the form of multiplication of the nuclei of the neurilemma and changes in the medullary sheaths. There were besides small capillary hæmorrhages in parts of the intracranial centres, one especially in the course of the abducens nerve in the pons Varolii. The alterations in the peripheral nerves were evidently primary and independent of those in the central organs. From the description it would appear that the changes in the nerve-fibres and neurilemma must certainly have preceded those in the nuclei of the oculo-motor nerves. Mendel concludes, and I think with justice, that the diphtheritic poison may attack not only nerve-sheaths and neurilemma, but also walls of blood-vessels, and that we have no right to assert that affection of one or the other of these tissues represents exclusively the pathology of diphtheritic paralysis.

It must be remembered that the cases in which disease of the spinal cord has been discovered have been of necessity fatal cases; and the question is, What is the pathology of the infinitely more numerous cases which not only recover, but recover without leaving a trace of any permanent change? I do not think that, with the clinical evidence before us, we are justified in saying that diphtheritic paralysis in its ordinary form, passing to complete recovery, is dependent upon an affection of the spinal cord. It is, in my opinion, more reasonable to conclude that in this disease we have usually to do with peripheral neuritis of very varying severity which in the mildest cases is probably represented by a mere transitory hyperæmia with effusion in the interstitial element.

Were such changes in the spinal cord as I have referred to ordinarily present, the complete recovery which is well known to be the rule would be impossible. The alterations in the number and structure of the ganglion cells would certainly be attended, as they are in cases of infantile paralysis, by more or less permanent paralysis and atrophy in corresponding nerve-districts. Nor should we find, as we do in many instances, the electrical reaction of the muscles remaining normal. Moreover, the sensory disturbance which is so often present in diphtheritic paralysis, cutaneous anæsthesia, tenderness of muscle and nerve-trunks, and darting pains, cannot possibly depend upon lesion of the ganglion cells of the anterior horns of the cord, whilst they point as strongly as possible to affection of nerve-fibres.

Practically, one very seldom indeed meets with examples of diphtheritic paralysis exhibiting such a near approach to the symptoms of typical interstitial neuritis as is so frequently seen in alcoholic paralysis. No doubt now and then a case is seen in which severe shooting pains are very prominent, another in which there is extreme and widespread loss of muscular power, or it may be that loss of cutaneous sensibility is strongly marked. So also here and there we may find one in which there remains a narrowly localised and permanent muscular atrophy.

I saw not long since, in consultation with Drs. Playfair and Semon, a case which illustrates the more rare consequences of diphtheritic paralysis.

The patient, a lady aged forty-two, was suffering from difficulties of articulation and occasionally of deglutition. She had been attacked eleven months previously with a diphtheritic sore-throat. Apparently she did not suffer badly, but kept her bed for ten days, during which there was at no time much fever. The sore-throat was followed very soon by paralysis of the soft palate, slight pain and unwieldiness of the tongue and weakness of the voice, and paresis of the upper extremities.

When I saw her (nearly a year having elapsed) the right half of the tongue was considerably atrophied, and when protruded inclined much to the patient's right. It quivered a good deal. The action of the lips was not nearly so lively as it should be, the result being that the lower half of the face was comparatively expressionless. I was not sure, likewise, that there was not some slight weakness of the right pterygoid muscle, for in opening her mouth and putting out the tongue, the jaw appeared to be carried somewhat over to her right. Sensation was quite normal. There was no tenderness on pressure over the points of emergence of the fifth nerve. The soft palate moved freely both on irritation and during phonation.

Dr. Semon reported that he found nothing wrong when the larynx was examined by the laryngoscope, though the character of the voice was rather peculiar and often changed into falsetto. The sterno-mastoids appeared equally strong and well developed. The patient complained of difficulty of articulation, especially in the use of the latter.

The tendon-reflexes were considerably exaggerated. This had not been the case, Dr. Semon said, two months previously, and he was disposed to refer the increase of reflex activity to the large quantity of strychnia which the patient had taken.

In this case the symptoms indicate lesion of the following nerves, or of their deep origin: the hypoglossal, facial, pterygoid branch of the inferior maxillary. Dr. Semon suggests that the curious character of the voice appears to point to some minute lesion of the spinal accessory. The limitation of the facial paresis to the muscles about the mouth points rather to lesion of a nerve than of its centre, and so does the weakness (if such there were) of the right pterygoid muscle, but it does not make this certain.

If we take a very common type of alcoholic paralysis and the most frequent form of diphtheritic paralysis, the contrast appears to be strongly marked. In the former we have notable paralysis of the extensor muscles of the extremities, with cutaneous tenderness or anæsthesia, pains of an agonising description, rapid muscular atrophy, and complete loss of faradic excitability. In the most common examples of diphtheritic paralysis there is weakness and ataxy rather than marked paralysis, a varying amount of cutaneous anæsthesia, pains either absent or of trifling character, no muscular atrophy, and electric reaction to induced currents either normal or but slightly reduced.

But alongside of these strikingly contrasted symptoms there are others which point more or less strongly to a kindred origin. In each case, for example, loss of patellar tendon-reflex is a very constant symptom, and tenderness of the muscles to palpation is likewise very generally marked in both. If, again, we pass from the strikingly contrasted forms to which I have just referred, we shall find cases more and more approaching a common character—cases of alcoholic paralysis with ataxic symptoms predominant, with pains but slightly marked or absent and only a moderate amount of anæsthesia; or, it may be, cases of diphtheritic paralysis without affection of accommodation or of the palate, and with ataxy rather than paralysis.   Moreover, in some rare cases of diphtheritic paralysis there may be exceedingly severe lancinating pains, great tenderness of the muscular masses, "dropped" hands and feet, followed by more or less general powerlessness of the upper and lower extremities, absent tendon-reflex, without affection of accommodation or of the velum palati—a train of symptoms, indeed, which is literally only distinguishable from those of alcoholic paralysis by the absence of any mental disorder.   It is hardly possible to doubt, from the mode in which the clinical characters of the two conditions thus approach each other and sometimes meet, that they depend essentially upon similar lesions varying in severity in different instances.

DIAGNOSIS.

From what has gone before it is evident that we may have to do, as regards peripheral neuritis, with either of several conditions : (1) There may be paralysis in the district innervated by a single nerve ; (2) or in that supplied by a plexus of mixed nerves ; (3) or the paralysis may be symmetrical and become more or less universal.

In the first case, if it be one of severity and the nerve concerned be a mixed one, there cannot be much difficulty in the diagnosis. There will be loss of power in the muscles supplied by a particular nerve, and in those only, with more or less atrophy of their substance and loss of electrical excitability, together with pain in the nerve-trunk, and abnormal sensation of some kind—numbness, formication, or anæsthesia of the skin in the district supplied by the nerve. If pressure be made upon the nerve-trunk, it may be found to be tender, and in certain cases will be felt to be enlarged. A moderate pressure is likely to give rise to a tingling sensation referred to the periphery of the nerve distribution. If the nerve in question, instead of being mixed, is purely motor in function, the evidence afforded by alterations of sensibility is wanting, and it is then often very difficult, and may be impossible, to decide whether the symptoms observed are due to changes in the trunk of the nerve, or in the nucleus of its central origin. A consideration of accompanying circumstances, together with the history of the attack, can alone afford help in deciding this point.

In the second case the association of sensory disturbance with motor paralysis would serve to distinguish lesion of a plexus from disease of the anterior grey matter of the cord.

As regards the third category, it is necessary to enter into more detail. The diagnosis of multiple neuritis is of great importance, and may present no small difficulty. It is evident that there are certain points in the symptomatology of this disease which enable us to narrow the question of diagnosis, so far at least as strongly-marked cases are concerned. The alteration in the electrical reactions of the muscles alone permits us to say at once that the disease is not dependent upon an intracranial lesion. We know that we have to do with paralysis of either spinal or peripheral origin.

In its mode of onset and progressive character there is a strong *primâ facie* resemblance between multiple neuritis and what is called the acute ascending paralysis of Landry. This disease (of which I know but little, save what I have read), beginning usually like multiple neuritis with numbness in the finger-ends and feet, is characterised by motor paralysis, which commonly first affects the lower extremities, and spreads upwards over the trunk and upper extremities. The circumstance that the distal portions of the extremities are first involved gives rise to a great resemblance to the aspect of multiple neuritis. But in acute ascending paralysis the general sensibility is said to be scarcely, if at all, affected; there is no notable atrophy of the muscles, and no change in their electrical reaction.

Weakness of the arms, with pain in one or other of them, accompanied by loss of power in the lower extremities, and anæsthesia extending to the upper ribs, may be caused by a tumour pressing upon the cord in the lower cervical region of the spinal cord. But in such a case we should expect to find more or less paralysis of the bladder, with a strong tendency to the occurrence of bedsores. The reflexes, deep and superficial, would be increased below the waist—the electrical excitability of the muscles of the lower extremities would be unchanged. There would be notable tendency to spasmodic contractions of the lower limbs.

The symptoms arising from Pott's disease of the spinal column when the affected vertebræ are in the cervical region are of course very comparable with those resulting from tumour in the same region. The symptoms of com-pression-lesion just enumerated will be found, and will alone suffice to distinguish from mul-tiple neuritis, even without the evidence afforded by the spinal deformity.

Where there is paraplegia from acute softening of the cord, the symptoms are far more
rapid than is the case in multiple neuritis. The
bladder is apt to be paralysed quite early in the
case—sometimes, indeed, a failure to pass urine
is the first symptom. Disorder of the bladder
and rectum are ordinarily as conspicuous in
this disease as they are altogether absent or
trifling in multiple neuritis. There is also constant tendency to destructive bedsores in the
sacral region and other parts, in striking contrast with the immunity from such lesions which
almost always obtains in multiple neuritis.

The diagnosis of multiple neuritis from spinal
lepto-meningitis is more complicated; for in the
latter there are shooting pains in the course
of nerve-trunks, with more or less hyperæsthesia
or anæsthesia in the districts of certain nerves,
together with impairment of muscular power.
But in meningitis we should find superadded to
these symptoms deep-seated pain in the back,
materially increased by movement, especially by
turning from side to side in bed. There would
probably also be rigidity of the spine, as well as
muscular spasms in the limbs. None of these
symptoms form part of the symptomatology of
multiple neuritis. Perhaps the greatest difficulty might be found in the differential dia-

gnosis from cervical pachymeningitis with extension of meningitis of the soft membranes through
the length of the cord. In this condition there
might easily be a more or less complete loss of
power of grasp, coupled with darting or burning
pains in the nerve-trunks of the arms, exactly
resembling those which are apt to occur in the
course of multiple neuritis. Moreover, there
would be atrophy of certain of the muscles of
the hands and arms, with loss of electrical excitability. It is not, indeed, surprising that
there should be this striking resemblance,
because when the cervical dura mater is thickened by inflammation (which necessarily involves
also the soft membranes below it), the lesion
attacks likewise both anterior and posterior
roots of the nerves and sets up neuritis in them.
But besides the symptoms in the upper extremities, the pressure on the cord in the cervical
region would occasion loss of power in the
lower extremities, combined with more or less
sensory and trophic disturbance in them according to the degree and extent of the inflammation
of the soft membranes. But in this condition,
again, we should expect to find stiffness of the
neck, with a good deal of pain in the back, increased by movement, and twitching in the
muscles of the extremities. Moreover, there

would be a great preponderance of symptoms referred to the *upper* extremities, whereas in multiple neuritis it is usually the lower extremities which suffer most.

In meningitis it is the roots or central ends of the nerves which are the subject of inflammatory change. The symptoms resulting will therefore be liable to appear equally developed in the various districts of distribution of the nerves, and will not be confined, as is so often characteristic of multiple neuritis, to the distal portions of these districts. The localisation of pronounced paralysis and sensory disorder in the periphery of the limbs points most strongly to multiple neuritis, but in certain cases a more or less extensive spread of the lesion upwards towards the centre may easily mask the significance of this sign.

Hæmorrhage into the substance of the cord or outside the membranes will, if extensive enough, occasion motor paralysis of limbs and trunk, with complete loss of cutaneous sensibility, muscular atrophy, as well as loss of electric excitability. The suddenness of the onset will alone suffice to distinguish such a case from one of multiple neuritis, to say nothing of the fact that paralysis of the bladder and bowel, together with strong tendency towards the occurrence of bedsores, would almost necessarily form part of the symptoms.

I have already referred to the resemblance to tabes dorsalis found at a certain stage of multiple neuritis. The patient is able to walk, but it is with an ataxic gait; the knee phenomenon is absent, and he complains very likely of sharp shooting pains. The alteration of electrical reaction in the muscles of the affected limbs will serve, when present, to distinguish the condition. But at a certain stage of recovery from multiple neuritis the muscles may be found to respond normally to induced currents. In these circumstances the difficulty of diagnosis on the part of anyone who had not watched the case in its antecedent stages would be very great. A little time, however, would clear up the obscurity, for the return of the knee-phenomenon would show that the case was not one of tabes.

The subacute atrophic spinal paralysis described by Duchenne is characterised by symptoms on the motor side closely resembling those of multiple neuritis, but it differs from the latter in presenting no affections of sensibility.

The differential diagnosis between multiple neuritis and acute anterior poliomyelitis may be very easy indeed, or so difficult as to give rise to considerable doubt. It appears to me that there are three principal points to be borne in mind : (1) In acute anterior poliomyelitis, what may be called the first stage—the stage of increasing intensity of symptoms—is usually much shorter than in progressive multiple neuritis, the paralytic symptoms far more complete, and the motor disturbance much more marked than the sensory symptoms, where these chance to be present. In multiple neuritis, on the contrary, such forms of the latter as " numbness," " deadness," and " pins and needles " are usually more prominent at first than the loss of power. (2) In acute anterior poliomyelitis groups of muscles functionally related are apt to be struck simultaneously with complete loss of power, whilst in progressive multiple neuritis the groups of muscles invaded by the disease are apt to be those in the district of distribution of various nerve-trunks rather than of plexuses. (3) In progressive multiple neuritis, severe enough to cause marked paralysis, you may expect to find distinct tenderness if you press upon the trunks of nerves where these are superficial. Sharp shooting pains in the course of peripheral nerves, lasting several days, furnish in an otherwise doubtful case conclusive evidence in favour of progressive multiple neuritis. This is the best diagnostic scheme which I can suggest, but I acknowledge that in certain cases it will not be sufficient for the required distinction. For it may chance, in a case of multiple neuritis, that the motor disturbance is exceptionally rapid, severe, and unaccompanied by sensory disturbance. It may be so extensive as to merge in one common powerlessness all the muscles of a limb, whether functionally or anatomically related. As regards the tenderness on pressing the nerve-trunks, I am not able to say whether this is constant in cases marked by motor symptoms chiefly or entirely, as well as in those characterised by striking pain and hyperalgesia. It is in the latter only that I have noted the symptom.

No doubt, if we take the example of a purely motor nerve—the portio dura,—in cases of neuritis of its trunk causing facial palsy, there is not any tenderness on pressure. But in a spinal mixed nerve, a nerve containing sensory as well as motor fibres, even when the symptoms point only to lesion of the motor nerve-fibres, it may well happen that there is over-excitability of the sensory fibres. But, as I have said, whether this is the case or not I do not know.

It may be asked whether, considering that in each case the lesion of the peripheral nerves is immediately the cause of the paralytic symptoms, it is not over-refining to try to determine to which category a particular example belongs. The importance lies in the fact that the prognosis is, on the whole, far more favorable in a case of multiple neuritis than in one of anterior poliomyelitis. The explanation of this cannot be better put than it has been done by Dr. Poore, who, in his selection from the works of Duchenne, reminds us of the " probability that the repair of motor cells once lost is impossible, whilst the repair of conducting fibres emanating from healthy motor cells need never be despaired of."

It will be remembered that typical cases of infantile paralysis are characterised by sudden onset, with or without fever, and paralysis, at first complete and extensive, gradually diminishing and becoming restricted to a greater or less number of muscles. It is characteristic that after the first day or so any change which takes place in the power of movement is a change for the better. The limbs do not become more paralysed; on the contrary, after a few weeks, or sometimes days, there is a gradual clearing off of the powerlessness as regards some of the limbs, one or more remaining perhaps unimproved, or the paralysis may remain limited to a few muscles in one limb. Many of the paralysed muscles lose their faradic excitability entirely within a week and rapidly waste. But although they fail to respond to the strongest induced currents they react to slow interruptions of the galvanic current. The nerves to the muscles lose their excitability to both forms of electrical excitation. Some of the muscles whose faradic excitability has been lowered but not lost, are not long in regaining the power of contraction to voluntary impulses. This rapid loss of faradic excitability is peculiar to the disease. In no other form of more or less generalised paralysis (which is not accompanied by marked sensory disturbance) do you find within a few days that the muscles fail to contract to induced currents. In this disease the pathological change is found to affect the anterior horns of grey matter in the spinal cord in which areas of softening are found, with disappearance of numerous multipolar ganglion cells, those that remain being degenerated and atrophied. The anterior roots of the nerves are atrophied. The lesion of the ganglionic cells has very much the same effect upon the motor nerve-fibre that would be produced by section of the nerve. Structural changes are produced throughout its ramifications.

Now, although the pathological changes in the anterior horn constitute the essential lesion in infantile paralysis (anterior poliomyelitis), yet it seems likely that at first the disease process may include more or less of the entire section of the spinal cord. In this way it has been sought to explain the circumstance that although sensory disturbance forms no part of the typical disease, and, indeed, is conspicuous by its absence, yet in some cases at the onset we find that sharp pains occur and tenderness of the limbs to the touch is complained of. It is the occasional occurrence of these symptoms in cases of infantile paralysis which renders it possible that cases of multiple neuritis may be mistaken for examples of anterior poliomyelitis. Some of us perhaps, and in this I include myself, have been too much inclined to mass under the name of anterior poliomyelitis all cases of paralysis occurring in children which have been characterised by sudden onset with loss of faradic excitability in the muscles, even when the amount of sensory disturbance has been greater than could be easily explained by diffusion of the pathological area over more or less of the whole section of the cord.

It is highly probable then that a certain

number of cases of so-called "infantile paralysis" are examples of multiple neuritis. I am much disposed to think that in the cases of "infantile paralysis" which make unexpectedly good recoveries after very long delay, the lesion may have been in the nerve trunks, and not in the anterior ganglia of the cord.

PROGNOSIS.

A very few words are necessary in reference to prognosis. In the early stage of multiple neuritis of non-alcoholic form, when the disease is spreading almost hourly so as to invade fresh nerve districts, the prognosis is necessarily an anxious one. The cardiac and respiratory apparatus may easily become involved, and death occur almost suddenly. But it is quite remarkable, as was seen in the two cases of my own which I have related, to what an extent the respiratory apparatus may become affected and recovery yet take place. When the disease seems to be no longer making fresh inroads, but, on the contrary, slight ameliorations begin to appear, a highly favorable result may generally be looked for. It is not so easy to speak as regards the alcoholic cases, as in them the brain also is always more or less involved in the disorder. But, as I have already remarked, my personal experience has decidedly disposed me to give a very favorable prognosis even in cases which are marked by extensive paralysis and muscular atrophy.

In diphtheritic paralysis the prognosis is distinctly favorable. It is probably through invasion of the pneumogastric that a fatal result now and then occurs, and for that reason serious modifications of the circulation, especially if accompanied by vomiting, should cause anxiety and care. So long as the knee-reflex is absent the patient should be looked upon as still an invalid, and not allowed to be incautious. This will obtain equally in other forms of multiple neuritis.

TREATMENT.

There are various degrees of severity shown by this disease, from a slight loss of power scarcely noticed by the patient, to a rapidly extending and complete paralysis, involving not only the nerves of the extremities and trunk, but also those belonging to the organs whose functions are essential to life.

The treatment of localised neuritis occurring in a person of gouty habit, and presumably dependent upon that exciting cause, should be in accordance with the customary methods adopted for acute gout. After the acute symptoms have subsided, good is sometimes experienced from the application of small flying blisters in the neighbourhood of the affected nerves, and comfort may be derived from the employment of the constant current.

In severe cases of progressive multiple neuritis it is advisable to place the patient at once upon a water-bed, although it will have been noted that in this disease there is not seen the same tendency to dangerous bedsores which is observed in certain lesions of the spinal cord. But I think that the water-bed supports the weak patient better than an ordinary couch; and this becomes of importance when, as sometimes happens, life is carried on with difficulty, owing to the nerves presiding over respiration, deglutition, and the heart's action becoming involved in the progress of the disease. In such cases as these the administration of food in easily assimilable form and quantity, and at very short intervals, is urgently required, and stimulants are often needed. During the first stage of the disease it will be advisable to administer iodide of sodium in all cases except examples of diphtheritic paralysis and those in which a syphilitic taint can be safely put aside. In the progress of the disease it may be found necessary to give the salt in increasing doses. In one of my cases the dose of iodide of potassium was increased from ten to sixty grains three times a day with evident advantage, and to this mercurial treatment was added. In a case of multiple neuritis with a distinct syphilitic history, I should now begin with mercurial inunctions, and also employ iodide of sodium at the same time. In cases of non-specific character, and especially where there is reason to think that exposure to cold and other causes of rheumatism have been present, it will be well to employ the salicylate of sodium, which in the hands of Leyden has apparently yielded favorable results. The dose and mode of use is like that for acute rheumatism. For the relief of pain a combination of morphia with Indian hemp and belladonna may be employed internally with advantage, and lint steeped in chloroform may be pressed for a minute or two on the seats of greatest suffering if the state of the skin admits of this. But very often, and especially in alcoholic cases, there is an amount of exqui-

site hyperæsthesia which renders it difficult to
apply any local remedy.   In such instances the
best thing is to envelop the tender limb in
cotton-wool, and cover this lightly with oil-silk.
When multiple neuritis has arisen in connection with the abuse of spirits, I am accustomed,
as a general rule, to withhold alcohol in any
form, and to depend entirely upon frequent
administration of food for the support of the
patient.   Nutrient enemata will sometimes be
required.   In such cases as these it is remarkable how rapidly the pains and hyperæsthesia
which have been the cause of intense suffering
to the patient cease.

It is very difficult to say how long a patient
suffering from multiple neuritis should be kept
strictly and absolutely at rest.   This should
certainly be done during the continuance of
pain or hyperæsthesia, and in case there is any
important elevation of temperature.   But when
it is evident that the process of regression or
repair has taken place to a considerable extent,
the patient should be allowed to get up and cau-
tiously try to move the muscles of the affected
limbs.   By slow and careful steps the effort at
voluntary movement may be increased.   At the
same time the galvanic current slowly interrupted should be applied to the muscles.

The motive for the application of the galvanic current may be thus explained. When you apply the two rheophores of an induced current machine to the skin covering a healthy muscle, and thereby obtain a contraction of muscular structure, that contraction is probably brought about by the momentary currents stimulating the intra-muscular nerve, and not by the action of the currents directly on the muscular fibre; so that when the conductivity of the nerve-fibres is lost by disease one result is that you fail to produce contraction of the muscle by applying the induced current to it. The essential element of the nerve-fibre has degenerated, but the muscular fibre has not (at least this obtains for a long time) suffered any important change. It is still, therefore, capable of being thrown into contraction by the making and breaking of a galvanic current which acts directly upon muscular tissue itself—not indirectly, like the induced current, through the medium of a nerve twig. Now it is probable that muscular structure may retain its power of contraction to this direct electrical stimulus for a very long time after it has lost its physiological connection with the anterior grey matter of the spinal cord. I have seen the property retained in these circumstances for several years

—in one case thirteen years—but I have also seen it lost altogether within a year. As the process of regeneration of nerve-fibre may require in severe cases a very long time for its completion, it is evidently important that the muscular structure should not by the time the nerve has recovered have itself undergone a degeneration from which there would be no return, and which would render it unable to respond by contraction to impulses arriving through the regenerated nerve. Physiologists have found that if muscle which has lost its nervous elements be artificially stimulated now and then to contraction, its irritability will continue. When left to itself, however, the irritability is apt to disappear and the muscular substance to undergo degeneration. There is, therefore, a good reason for the use of slow interruptions of the constant current in these cases.

There are some grounds for believing also that faradisation with the wire-brush upon the dry skin may be employed with advantage. As you are aware, this means was introduced by Duchenne for the treatment of anæsthesia. There can be no doubt of its service in anæsthesia, and by inference we may perhaps reasonably conclude that it may help by its action on the sensory fibres of the nerve towards the restoration of function in the lesion of nerves under consideration. Massage is also useful in this stage, and in this I would include passive movements by the operator, as well as active movements against resistance on the part of the patient. In the contracted state of limbs which occasionally results, the contracture being due to unbalanced muscular antagonism, division of a tendon may sometimes be adopted with advantage. Considerable patience should be employed before proceeding to this measure, as I have known contractures, which were to all appearance hopelessly permanent, yield, without operation, to assiduous massage combined with active and passive movements. Along with the contracture of the muscles, it will sometimes be found that adhesions have taken place in some of the joints, owing to disuse. These should be forcibly broken down. The aim generally should be to disengage muscles from obstructions to their movement, and to encourage their growth and functional activity by various kinds of physiological stimuli.

INDEX

A

Alcoholic paralysis, 70—93, 105, 106; compared with diphtheritic paralysis, 120; marked occurrence of pain in, 82; and of muscular atrophy, 82

Ataxy in alcoholic paralysis, 73, 121; in diphtheritic paralysis, 110, 112, 113, 121

Atrophy of muscle, 12, 28, 50, 53, 59, 63, 69, 72, 75, 82

B

Bedsores, 61, 62

Beriberi, 5, 64; Hirsch on, 64; Palm on, 65; Scheube on, 67; Harada on, 67; Pierson on, 68

Broadbent on alcoholic paralysis, 77

C

Central disease simulated by peripheral neuritis, 4, 23, 24

Cervical pachymeningitis, 127

Churton on a case of ataxy, 89, 90

Contracture, 63; treatment of, 142

D

Degeneration, reaction of, 9

Déjerine on peripheral neuro-tabes, 91

Dengue, peripheral neuritis after, 102

S

Scheube on Beriberi, 67

Semon, Dr., on a case of diphtheritic paralysis, 119

Sensory symptoms in multiple neuritis, 61

Skin, trophic changes in, 18, 19

Softening of spinal cord, 126

Spinal cord, tumour of, 125; softening of, 126; hæmorrhage into, 128

Stewart, Dr. Grainger, on multiple neuritis, 49

Syphilitic antecedents in multiple neuritis, 43; treatment of, 138

T

Tabes dorsalis, 88, 113; differentiated from multiple neuritis, 113, 129

Tenderness of nerves and muscles, 109, 111, 114, 118, 121

Tingling of skin, 86

Tumour of cord, 125

Treatment, 137

Trophic changes in skin, 18, 19

U

Use of galvanism in multiple neuritis, 140

W

Wallerian degeneration, 8

Watteville, Dr. de, on electrical resistance of the skin, 32; on lead paralysis, 55

Wilks on alcoholic paralysis, 78

PRINTED BY J. E. ADLARD, BARTHOLOMEW CLOSE.

www.ingramcontent.com/pod-product-compliance
Lightning Source LLC
Chambersburg PA
CBHW021814190326
41518CB00007B/592